A Draught of the South Land

A Draught of the South Land

Mapping New Zealand from Tasman to Cook

Paul Moon

The Lutterworth Press

The Lutterworth Press

P.O. Box 60
Cambridge
CB1 2NT
United Kingdom

www.lutterworth.com.
publishing@lutterworth.com

Hardback ISBN: 978 0 7188 9721 5
Paperback ISBN: 978 0 7188 9720 8
PDF ISBN: 978 0 7188 9719 2
ePub ISBN: 978 0 7188 9722 2

British Library Cataloguing in Publication Data
A record is available from the British Library

First published by The Lutterworth Press, 2023

Contents

List of Illustrations

Introduction

> Sky full of ships, bay full of town,
> A port of waters jellied brown:
> Such is the world no tide may stir,
> Sealed by the great cartographer.
> O, could he but clap up like this
> My decomposed metropolis,
> Those other countries of the mind,
> So tousled, dark and undefined!
>
> Kenneth Slessor, *The Atlas*[1]

Now when I was a little chap I had a passion for maps. I would look for hours at South America, or Africa, or Australia, and lose myself in all the glories of exploration. At that time there were many blank spaces on the earth, and when I saw one that looked particularly inviting on a map (but they all look that) I would put my finger on it and say, 'When I grow up I will go there.' ... Other places were scattered about the Equator, and in every sort of latitude all over the two hemispheres. I have been in some of them, and ... well, we won't talk about that. But there was one yet – the biggest, the most blank, so to speak – that I had a hankering after. True, by this time it was not a blank space any more. It had got filled since my boyhood with rivers and lakes and names. It had ceased to be a blank

1. K. Slessor, 'The Atlas', in A.J. Haft, 'Imagining Space and Time in Kenneth Slessor's "Dutch Seacoast" and Joan Blaeu's Town Atlas of The Netherlands: Maps and Mapping in Kenneth Slessor's Poetic Sequence *The Atlas*, Part Three', *Cartographic Perspectives* 74 (2013), pp. 41-42.

> space of delightful mystery – a white patch for a boy to dream gloriously over. It had become a place of darkness.
>
> Joseph Conrad, *Heart of Darkness*[2]

For nearly 130 years after 1642, when Abel Tasman became the first European to sight and record New Zealand, maps containing fragments of information about New Zealand ricocheted around the world. Roughly sketched coastlines, pen-and-ink topographical silhouettes, a stockpile of names (the first accomplice of colonisation) attached to locations, and a mass of coordinates, measurements, calibrations and estimations were all copied, refined, translated and augmented as they spread among Europe's imperial powers and the continent's growing reading classes.

These two-dimensional representations were fleshed out by the copious commentaries that supplemented them and, as the number of maps multiplied, their form evolved. Monochromatic engravings began to blush with colour, quilled cartography gave way to printed text and images, and what were once solitary maps – buried deep in trading companies' archives – now appeared in mass-produced publications, in several languages, and fed the imaginations of new generations of explorers.

In the seventeenth and eighteenth centuries, intelligence about imperial discoveries ran through various arteries. From the heavily clotted official channels to the fast-flowing informal routes, cartographical information continued to circulate, despite occasional efforts to restrict its movement. Between the gossipy exchanges in colonial outposts, the map merchants in Europe churning out countless copies of distant territories, the agents of trading companies, and scholars keen to decipher the world's uncharted extremities, there was little prospect of maps in the seventeenth and eighteenth centuries ever being kept secret for long. The forces in favour of the democratisation of knowledge about the earth's territorial arrangements were simply too strong.

The specific focus on maps as part of the history of exploration, discovery and colonisation is a relatively recent field of enquiry. Roughly speaking, up until the mid-twentieth century, maps tended to be the preserve of specialist librarians and antiquarian collectors, for whom the aesthetic or suggestive qualities of these items was of prime importance. In the post-war period, though, there was a sea change in attitudes towards

2. J. Conrad, *Heart of Darkness* (Oxford: Oxford University Press, 2002), p. 108.

cartography, with the study of maps drawing on a range of academic disciplines and contributing substantially to understanding the ways in which Europe penetrated and perceived the world beyond its borders.[3] Previously, maps were generally seen simply as documents that displayed places and the spaces between them, and not much more than that. Their historical value tended to be confined to exercises in comparison as a means of documenting progress in cartographical technique and accuracy over time. In recent decades, however, such views of maps have fallen sharply out of favour. Instead of being works innocently depicting distances between territories, cartography has come to be portrayed as the advance guard to imperial assertions of cultural, political and geographic dominance,[4] exercising a form of centrifugal force on the way that the world outside Europe was portrayed and regarded. Both views – the naïve and the nefarious – represent opposing poles in the perception of maps. At different times, and in different hands, maps have served a range of overlapping functions and were never either wholly independent of or fully complicit with the much greater political, economic and cultural forces from which they emerged.

In the seventeenth and eighteenth centuries, maps were also an integral part of the flow of intelligence between European nations. The nature of states' involvement in the international arena was shaped in part on the intelligence they had at their disposal. However, paradoxically, in order to have access to this intelligence in the first place, they needed to be actively involved in the international realm.[5] During Europe's Age of Discovery (from roughly the sixteenth to the eighteenth centuries),[6] this balance tilted strongly in favour of intelligence gathering as the

---

3. M.H. Edney, 'Putting "Cartography" into the History of Cartography: Arthur H. Robinson, David Woodward, and the Creation of a Discipline', *Cartographic Perspectives* 51 (2005), pp. 14-15.
4. P.M. Smith, 'Mapping Australasia', *History Compass* 7, no. 4 (2009), p. 1099.
5. See, for example, M. Thomas, 'Colonial States as Intelligence States: Security Policing and the Limits of Colonial Rule in France's Muslim Territories, 1920-40', *Journal of Strategic Studies* 28, no. 6 (2005), 1033-60; K. Raj, 'Colonial Encounters and the Forging of New Knowledge and National Identities: Great Britain and India, 1760-1850', *Osiris* 15 (2000), 119-34; D. Killingray, 'The Maintenance of Law and Order in British Colonial Africa', *African Affairs* 85, no. 340 (1986), 411-37; P. Murphy, 'Creating a Commonwealth Intelligence Culture: The View from Central Africa 1945-1965', *Intelligence and National Security* 17, no. 3 (2002), 131-62.
6. J.H. Parry, *The Age of Reconnaissance: Discovery, Exploration and Settlement, 1450-1650* (London: Phoenix Press, 1962), p. i.

precursor for any decisions being made on foreign policy. This resulted in those nations which possessed the financial and technological means undertaking expeditions to seek out territories that were either unknown, unmapped or 'unclaimed'[7] by other European powers.[8] Yet, as the experience of New Zealand's emergence in maps reveals, the process was almost autonomous of any direct state control. Yes, plans were devised, capital was raised and orders issued for exploratory expeditions. However, once the intelligence had been gathered, it tended to take on a life of its own and proved beyond the ability of any single state in this period to control it fully. There were too many informal networks, too many places in the intelligence border where breaches could occur, too many entrepreneurs concerned with personal profit rather than state policy, and too many instances of institutional inertia and incompetence to maintain the type of restraint over intelligence that would be exercised with more success in later centuries.

How the shape of New Zealand came to appear in map form encompasses these trends and relationships. There is a risk, though, that the whole process comes to be seen as little more than the workings of anonymous bureaucracies, functioning in some great Europe-wide institutional ecosystem of clerks and cartographers. What this work shows, in contrast, is that the means by which the country came to be mapped was dependent on a range of individuals – many of whom were extraordinarily able in their respective fields – and often occurred in spite of state involvement, rather than because of it. It can be almost instinctive to attribute European mapmaking in this era to government directives and policies, but this would be to overestimate their influence. Ultimately, the image of New Zealand came into view on maps as a result of 'the tremendous shaping, form-creating force working from within' the minds of particular individuals.[9] Of course, no cartographer or explorer was an island, separate from the nations in which they lived and worked. However, while politicians and policy-makers endeavoured

7. See, for example, A. Fitzmaurice, 'The Genealogy of Terra Nullius', *Australian Historical Studies* 38, no. 129 (2007), 1-15.
8. C. Bayly, *Empire and Information: Intelligence Gathering and Social Communication in India, 1780-1870* (Cambridge: Cambridge University Press, 1996), pp. 2-3. Bayly's concept of 'information order' is derived from M. Castells, *The Informational City: Information, Technology, Economic Restructuring, and the Urban-Regional Process* (Oxford: Basil Blackwell, 1989).
9. F. Nietzsche, *The Will to Power* (New York: Vintage Books, 1968), p. 344.

to put in place schemes for their countries' overseas mapmaking expeditions, those involved in executing them were summoned as much by their own will and often uncompromising brilliance, and it was precisely from this volatile dynamic that the outline of New Zealand eventually came into view.

Note on Terminology

The terms 'chart' and 'map' generally have a distinct meaning. The former is normally used by sailors to navigate routes on bodies of water and may display features such as tides, currents and coasts, while the latter tend to focus on the layout of terrestrial features. However, in the centuries that this book surveys, the terms have been used in such a variety of ways – and often interchangeably – that the precise distinction is not depended on here. Instead, both terms are used to describe the representation of aspects of a location on paper (or a globe).

Chapter 1

Cartography and the Age of Discovery

Hessel Gerritsz and His Map of Batavia

Although he was highly prolific, the quantity of Hessel Gerritsz's output tended to dip from around November to March each year as the dim winter daylight placed too much strain on his eyes to craft the minute details that his clients required on their maps. Still, though, the demand for his services in the 1620s and early 1630s was undiminished. Thus, as the shopkeepers, city officials, clergy, pick-pockets and even the prostitutes began to desert Amsterdam's streets in the chilly darkening mid-afternoons each year during this period, Gerritsz would roll up his maps, sheathe them in a specially-designed calfskin satchel to protect them from the elements, and make his way along the cobblestone-lined Kloveniersburgwal canal to the imposing headquarters of his principal employer, the Dutch East India Company (identified by its Dutch acronym VOC – Verenigde Oostindische Compagnie).[1] When this monumental late-Renaissance edifice was completed in 1606, it was one of the largest buildings in Amsterdam and embodied in architectural form the financial muscle that the company flexed. While he waited for the main door at the entrance of the company's offices to be opened, across from the inky canal, Gerritsz would be able to see the outline of the neat row of tall, narrow baroque houses on the eastern bank, posing elegantly against the fading light and massing black clouds – a view that has survived to the present day.

1. The world's first multinational firm owned by shareholders.

After depositing his work, usually with one of the company's directors,[2] and having discussed his current projects and possible new commissions,[3] Gerritsz would head homeward, with only the flickering lanterns outside the doors of a few houses providing faint spots of iridescence along the way. As he walked back, he would hug his *huik* – a heavy hooded cloak – closer to his body to shield himself as much as he could from the icy air that funnelled through the Markermeer from the North Sea. It was a route he had taken for more than a decade since he had been contracted by the Dutch East India Company to produce and improve maps for:

> all places, regions, islands, and harbours relevant for the Asian navigation ... all logbooks that he already received and which he will receive from now on through the directors, will have to be stored in the East India House [in Amsterdam]; he will keep a complete catalogue of the logbooks ... he will correct the standard charts only after the directors have approved the corrections; on his death his widow or heirs will hand over all the papers in his possession.[4]

Tucked in at the end of the agreement that Gerritsz had with the company was the requirement that he would 'publish nothing without the permission of the directors ... [and] he should observe secrecy about his work'.[5] Breaching this would incur a fine of 6,000 guilders (the modern equivalent of just under £200,000). Such a heavy penalty was an indication of how commercially valuable maps were becoming to the company, particularly when its commercial empire was in fierce competition in this era with those of Portugal, England and Spain.

In addition to this contract, Gerritsz undertook commissions for the West India Company, the Dutch Admiralty of Amsterdam and other agencies if the work was available. At times, he collaborated with Willem Janszoon Blaeu, a cartographer ten years his senior, who specialised in atlases[6] and who passed on many of his cartographical

---

2. There were seventeen in total and they were known as the Heren XVII – the Gentlemen Seventeen.
3. J. Keuning, 'Hessel Gerritsz', *Imago Mundi* 6, no. 1 (1949), p. 56.
4. P. van Dam, *Beschryvinge van de Oostindische Compagnie*, ed. F.W. Stapel (The Hague: Martinus Nijhoff, 1927-54), Vol. 1, pp. 414-15.
5. Ibid., p. 415.
6. As an example, see W.J. Blaeu, *Atlantis majoris appendix sive pars altera continens geographicas tabulas diversarum orbis regionum et provinciarum octoginta* (Amsterdam, Willem Janszoon Blaeu, 1630).

skills to Gerritsz.[7] These were not the only mapmakers, however, who were employed by the trading companies in Amsterdam in this period and there was consequently a degree of rivalry among the studios[8] over which could produce the most dependable charts as promptly as possible for their wealthy clients.

In 1613, Gerritsz had attempted to further his chosen career as he neared the end of his apprenticeship by publishing a small book containing just one map. It was a copy of an English chart produced by John Daniels around 1611, which had made its way to the Netherlands after an English navigator – Allen Sallowes – deserted his vessel to escape debts he owed in England and befriended a group of Dutch sailors, to whom the map was passed on.[9] Instead of being a compendium of maps and charts, as might have been expected of an up-and-coming cartographer, Gerritsz's book was a scientific treatise on a Norwegian archipelago in the Arctic Ocean then known as Spitsbergen.[10] He saturated his text with geographical coordinates and descriptions of the location's terrain, coast and resources – the sort of information that he hoped would convince the Dutch East India Company of his abilities. Moreover, just in case this was not quite sufficient to arouse the company's interest, he deliberately seasoned the book's subtitle with an anti-English flavour (referring to the 'annoyances which the Whalers, Basque, Dutch, and Flemish, Have endured at the Hands of the English').[11] The ploy was successful and, with his loyalties and capabilities thus demonstrated, he soon entered into the circle of cartographers contracted by the company to map its expansion.

By 1630, Gerritsz was one of the greatest mapmakers in Europe. From his Amsterdam workshop – which was crammed with paints, palettes, props, piles of papers, easels, benches, books, works-in-progress and all the other bric-a-brac of his profession – this 49-year-old cartographer,

7. G. Schilder, 'Organization and Evolution of the Dutch East India Company's Hydrographic Office in the Seventeenth Century', *Imago Mundi* 28, no. 1 (1976), p. 62.
8. K. Zandvliet, *De groote waereld in't kleen geschildert: Nederlandse kartografie tussen de middeleeuwen en de industriële revolutie* (Alphen aan den Rijn: Canaletto, 1985), pp. 177-80.
9. M. Conway, 'The Cartography of Spitsbergen', *Geographical Journal* 21, no. 6 (1903), p. 638.
10. It is now known as the Svalbard archipelago.
11. H. Gerritsz, *History of the Country Called Spitsbergen: Its Discovery, Its Situation, Its Animals* [Amsterdam, 1613], trans. (London: British Museum, 1902), p. 11.

artist, publisher, surveyor and engraver produced territorial maps and maritime charts to a standard that was the equal of anything being created at that time.[12]

* * *

In almost every sense, Amsterdam was a world away from Batavia (now Jakarta) – the muggy tropical trading post that was at the farthest reach of Dutch mercantilism in the early seventeenth century and that bordered what was for Europeans in this era the edge of the known world. It was a map of Batavia produced by Gerritsz (based on a preliminary version sketched the previous year by the cartographer Jacob Cornelisz van Cuyck)[13] that he delivered to the company's offices on Kloveniersburgwal one afternoon in 1630. This map – created when urban planning was as much a capricious art as a pedestrian science – was comprised of a series of streets laid out in rigid vertical and horizontal lines (in keeping with the Roman grid system in which he had been trained),[14] with a serpentine waterway curving along part of its periphery being the only significant concession to the landscape. This was not the layout of an existing city, though, but one that officials of the Dutch East India Company hoped in the 1630s would sprout from the trading post they had planted in Batavia in 1619. The imagined form of the settlement fleshed out in Gerritsz's map was comprised of a fort, numerous commercial buildings, houses, a moat, city walls, canals, agricultural land and, of course, for this fiercely Protestant imperial power, churches.[15] This minor work of commercial, colonial and cultural ambition is as useful a point as any from which a chain of cartographical events can be traced, leading eventually to the creation of the first map of New Zealand.

---

12. G. Schilder, 'New Cartographical Contributions to the Coastal Exploration of Australia in the Course of the Seventeenth Century', *Imago Mundi* 26, no. 1 (1972), p. 41; K. Zandvliet, 'Mapping the Dutch World Overseas in the Seventeenth Century', in D. Woodward (ed.), *The History of Cartography: Volume 3 (Part 2): Cartography in the European Renaissance* (Chicago: University of Chicago Press, 2007), p. 1437.
13. J.C. van Cuyck, *Batavia in 1629 tijdens de Mataramse belegering*, General State Archives of The Hague, Maps and Drawings Department, collection VEL, inv.nr. 1179 A.
14. M.L. Kehoe, 'Dutch Batavia: Exposing the Hierarchy of the Dutch Colonial City', *Journal of Historians of Netherlandish Art* 7, no. 1 (2015), 1-35.
15. H. Gerritsz, *Plattegrond van Batavia en omstreken*, National Archives of the Netherlands, NL-HaNA_4.VEL_1179B; Kehoe, 'Dutch Batavia', p. 6.

Mercator's Boost to Explorers

Exploration has always been an appetite that grows with the eating. Certainly, the more that the Dutch charted the outer stretches of their mercantile empire – starting from the mid-sixteenth century – the keener they were to discover what opportunities lay beyond. Maps were a vital piece in the apparatus of this process and, at the time, represented a peculiar concoction of the mechanics of cartography and the exercise of imagination.

The idea that something existed past that hazy point where the ocean vanishes into the horizon had been part of the architecture of Western thought for millennia. Herodotus' fantastical account (written around 430 BC) of the Phoenician exploration of Libya, for example, emphasised the exoticism of territories beyond the known world.[16] Nearly a century later, Plato's story about the fictional country of Atlantis became the trope for the notion of a significant landmass – in this case 'larger than Libya and Asia'[17] – that was situated just outside the perimeter of current geographical knowledge.[18] Perceptions of the world in this era remained a rich mix of fact and fantasy, with uncharted regions 'a blank space of delightful mystery ... to dream gloriously over'.[19] However, just as nature abhors a vacuum, so do inquisitive minds get frustrated by the unknown. It was therefore inevitable that the inundations of speculation that had submerged ideas about the world's physiography would gradually begin to recede in the wake of a more disciplined approach to mapmaking.

It was Claudius Ptolemy – the great Egyptian astronomer, geographer, and mathematician[20] – who, around AD 150, undertook one of the

---

16. Herodotus, *'On Libya', from The Histories, c. 430 BC*, text provided by Prof. J.S. Arkenberg of California State University, Fullerton, Book 4, 42; available online at the Ancient History Sourcebook, History Department, Fordham University, New York (1998).
17. Plato, *Timaeus and Critias*, trans. R. Waterfield (Oxford: Oxford University Press, 2008), p. 105.
18. D. Clay, 'Plato's Atlantis: The Anatomy of a Fiction', *Proceedings of the Boston Area Colloquium in Ancient Philosophy* 15, no. 1 (1999), pp. 3-4; M. Edmond, *Zone of the Marvellous: In Search of the Antipodes* (Auckland: Auckland University Press, 2009), p. 23. These stories, in turn, were preceded by discovery narratives from Mesopotamia and Babylon. As an example, see N.K. Sanders (trans.), *The Epic of Gilgamesh* (London: Penguin, 2006).
19. Conrad, *Heart of Darkness*, p. 5.
20. H. Barnard, 'Maps and Mapmaking in Ancient Egypt', in H. Selin (ed.), *Encyclopaedia of the History of Science, Technology, and Medicine in Non-Western Cultures* (Berlin: Springer-Verlag, 2008), pp. 1273-76.

first serious efforts to produce a map of the world as it was known to Greco-Roman civilisation. Instead of embroidering the work with elaborately-stitched speculation about fictional continents full of fanciful creatures, Ptolemy's map was based solidly on the more dependable fabric of science. For the first time, longitudinal and latitudinal lines, along with astronomical observations, were used to give dimensions and anything that could not be verified was omitted. The work in which Ptolemy's maps appeared was titled *Geographia*,[21] and its careful cartography was accompanied by a detailed text which remained central to Europe's understanding of the world for more than a thousand years after it was written. Ptolemy's ambition was to represent reality rather than replace it and, in the following centuries, *Geographia* served as the archetype for academics concerned with mapping the physical form of the world. What few copies of it existed themselves became part of geography, travelling through Europe and the Middle East[22] in sprawling networks of knowledge. Somewhere en route, though, the unthinkable happened: the last remaining parchment of Ptolemy's original map was lost. Copies of it had been made, but over time, these too had disappeared into the oblivion of some remote cloistered library or monastic cell that was eventually abandoned, or perished in fires or floods, or at the hands of thieves looking for more easily tradable treasures.

What followed was one of the great acts of scholarly reincarnation. Around 1295, a Byzantine monk, Maximus Planudes, sought out a copy of *Geographia* and managed to obtain one, but with the maps missing.[23] He then began the painstaking process of reconstructing Ptolemy's map from the coordinates and other details contained in the surviving text.[24] The result was a map of the world that had not been seen possibly for many hundreds of years. Within a century, the 'new' Ptolemy map was migrating westward, reaching Florence in 1397, where *Geographia* was translated into Latin and then other languages.

---

21. J.L. Berggren and A. Jones, *Ptolemy's Geography: An Annotated Translation of the Theoretical Chapters* (Princeton: Princeton University Press, 2001). It was partly reliant on the work of a near contemporary, Marinus of Tyre.
22. The tenth-century Arabic writer, al-Mas'ūdī, referred to the content of *Geographia*'s maps. N. Levtzion, 'Ibn-Hawqal, the Cheque, and Awdaghost', *Journal of African History* 9, no. 2 (1968), 223-33.
23. F. Pontani, 'The World on a Fingernail: An Unknown Byzantine Map, Planudes, and Ptolemy', *Traditio* 65 (2010), p. 190.
24. A. Diller, 'The Oldest Manuscripts of Ptolemaic Maps', *Transactions and Proceedings of the American Philological Association* 71 (1940), 62-67.

The pace at which the new versions of Ptolemy's work spread throughout Europe quickened dramatically from the 1470s onwards when printed editions began to be published in Rome (and later elsewhere). Some volumes contained coloured images of the maps, enabling readers to see – in the most vivid tones that technology offered – the seas, deserts, mountain ranges, rivers and landmasses that made up the world. Sagging along the bottom of Ptolemy's reconstructed map was a red latitudinal line (incorrectly labelled the Tropic of Capricorn), which served as the boundary of what the West knew about the world in this era. Anything beyond that red line was 'outside the circuit of civilisation'[25] – a moral and geographical blank canvas where lay all manner of possibilities and pitfalls.[26]

By the fifteenth century, with *Geographia* circulating widely throughout Europe, the whiff of opportunity reached the flared nostrils of traders, clerics, military leaders and imperial-minded politicians, all of whom could almost smell the prospects for furthering their respective ambitions in new territories, and, although temporal interests tended to dominate plans to probe those areas that were still unmapped, there was often a religious dimension hovering over most of the exploratory expeditions that began to get under way from this time. Voyages of discovery were, in one sense, a type of maritime pilgrimage[27] and, for those Catholic nations engaged in the pursuit of new lands, the blessing of the Church became a form of spiritual sanction for what was otherwise a very secular undertaking. Mapmaking consequently took on more pronounced religious functions at times. In one reasonably representative example, attributed to Amerigo Vespucci, the early sixteenth-century Florentine explorer, the author noted during his voyages 'the marvellous works wrought by that sublime creator of all things, our God' and prayed that, through his acts of discovery, 'the immense work of almighty God, partly unknown to the ancients, but known to us, may be understood'.[28] Maps could thus be construed as acts of divine revelation – making known parts of Creation that were previously concealed.

---

25. D.H. Lawrence, *Sea and Sardinia* (New York: Thomas Seltzer, 1921), p. 3.
26. J. Axtell, 'Europeans, Indians, and the Age of Discovery in American History Textbooks', *American Historical Review* 92, no. 3 (1987), p. 631.
27. J. Eade and M.J. Sallnow (eds), *Contesting the Sacred: The Anthropology of Christian Pilgrimage* (London: Routledge, 1991), p. x.
28. A. Vespucci, *Mundus Novus: Letter to Lorenzo Pietro di Medici* [Vienna, 1504], trans. G.T. Northup (Princeton: Princeton University Press, 1916), p. 11.

For a while, science was happy to cohabit with dogma – each was seen simply as a different side of the same coin. However, whilst dogma was encased by the edicts of the Church, science was becoming freer to flourish and, in some parts of Europe by the end of the sixteenth century, was increasingly unfettered by papal restrictions. Cartography fell comfortably into the category of a science and was bound by the emerging enlightenment principles of experimentation, observation and verification. Unlike fields such as chemistry or mathematics, though, cartography also had one foot in the world of politics and commerce. Maps outlined the parameters of power and charted routes to profits. They were a means of quantifying terrain, which was the necessary precursor to claiming some sort of interest in that terrain, and possessed the power to define a state sometimes even before the inhabitants of that territory knew of that state's existence.[29] The demarcation of lands on maps even had a quasi-legal dimension, as though they were a silent assertion of property rights. Indeed, in the absence of the bureaucratic apparatus of land titles and written registration of ownership in many territories outside Europe, maps sufficed as an interim substitute (while showing no regard for alternative landownership concepts that might have prevailed in some of the affected regions).

What maps omitted was often just as important as what they depicted. In the seventeenth century, the cultural, political and social landscape of an area was usually absent from maps. Instead, when trading-company directors or state officials laid out a chart on a table and hunched over it to examine its details, they were looking at a two-dimensional outline of territories, with perhaps a few topographical pointers added, but little more. The lack of political mapping (such as demarking the boundaries of indigenous communities) was itself a political gesture, based on the presumption that the only sort of political power that mattered was the European variety.

Maps were also vital in laying the groundwork for the establishment of colonial administrations in some territories.[30] A fledgling colonial presence in a remote region needed to have some sense of the reach of its authority, and existing boundaries on maps could be useful (though not necessarily binding) in determining where colonial jurisdiction applied.

---

29. M.H. Edney, 'The Irony of Imperial Mapping', in J.R. Akerman (ed.), *The Imperial Map: Cartography and the Mastery of Empire* (Chicago: University of Chicago Press, 2009), p. 13.
30. J.C. Stone, 'Imperialism, Colonialism and Cartography', *Transactions of the Institute of British Geographers* 13, no. 1 (1988), p. 61.

There was something slightly predatory in this whole process, though. Cartography laid out the contours of colonisation and was often used by nations to signal where their imperial intentions lay, or that their grip on a particular territory was tightening. Moreover, for the regions being colonised, maps imposed a European construction of geographical space. For the Dutch in the seventeenth century, maps portrayed a specific economic geography and the routes to commercial enlargement. A territory roughly outlined was usually an indication of its low worth or priority, whereas more economically desirable locations were often densely detailed.

Maps were therefore hardly neutral or purely objective exercises in geographical depiction[31] but, instead, were part of the process of drawing in remote and formerly unknown territories into the consciousness of Europe. Once imbedded in the collective mind of the colonising nation, maps imbued those remote territories with a history and sense of meaning that they supposedly previously lacked. Maps were the first step in making the exotic and obscure (conjured up in the adventurous-sounding yet literal term 'uncharted territory') familiar, comprehensible and even slightly mundane. To be mapped by Europe in this era was to be put on notice that, at some point, Europe would carry out its designs on your territory. There simply could be no colonisation or commerce without cartography. These elegantly penned and generously coloured charts were not so much illustrations as manifestos, and those involved in Amsterdam's mapmaking industry in the seventeenth century understood this very well. After all, it was the schemes of colonists and traders which kept them in business. The sorts of sophisticated maps being produced in the Netherlands in the early seventeenth century were themselves possible in large part because of corresponding advances in fields such as ship design, surveying techniques, navigational technology and improved communications. These factors, coupled with the fact that profitable trading companies could afford to invest in high-quality cartography, resulted in Amsterdam becoming one of the most important mapmaking centres in Europe.

Images of the world, as it was pictured in the sixteenth century, had been circulating throughout Europe in this period. In the course of their travels, copies of these maps were made, errors reproduced and

31. R.B. Craib, 'Cartography and Power in the Conquest and Creation of New Spain', *Latin American Research Review* 35, no. 1 (2000), p. 8; B. Hooker, 'New Light on Jodocus Hondius' Great World Mercator Map of 1598', *Geographical Journal* 159, no. 1 (1993), 45-50.

compounded, and misconceptions were converted into certainties. This sort of laxity was permissible with maps designed to entertain those armchair explorers who were wealthy enough to afford them and who treated them largely as (expensive) curios. However, merchants needed something more dependable if they were to consider venturing to new locations to expand their businesses. It was the mapmaker and geographer Gerardus Mercator (born in modern-day Belgium), who disrupted the prevailing cartographic conventions with his publication in 1569 of a heavily-annotated atlas of the territories of the globe known to Europe. It was the first attempt at creating a conformal image of the world, based on a cylindrical projection, which enabled the Earth's meridians and latitudes to be represented on the flat surface of a map[32] – a technique that was of immediate benefit to maritime navigation. Gerardus' son, Rumoldus, inherited his father's devotion to cartographic accuracy and, in 1595, published a revised version of the atlas. Rumoldus died in 1599 and, five years later, the copper plates of the atlas he and his father had made were purchased by his brother-in-law, the Amsterdam-based cartographer, Jodocus Hondius. Hondius was famous for the accuracy and artistry of his maps and subsequently published several editions of what became known as the Mercator-Hondius Atlas,[33] which were works of renewal rather than recycling. By the early 1600s, Amsterdam had emerged as one of the centres of European mapmaking and, unlike cartographic works published in some other countries, the Dutch made an effort to translate many of their publications (especially into French), which immediately opened them up to new markets and had the effect of making the map business in Amsterdam more lucrative still.

From the 1630s, a new and even more industrious generation of cartographers had established itself in Amsterdam. As popular demand for their atlases grew, there was something of a publishing arms race, with a rush to produce larger works containing more maps. Blaeu released his *Atlantis Appendix* in 1630 – a volume comprising of 60 maps – and then reissued the work the following year with an additional

32. L. Farrauto and P. Ciuccarelli, 'The image of the divided city through maps: The territory without territory' (Indaco Department, Politecnico di Milano, January 2021), p. 12; R.H. Charlier and C. Charlier, 'Lowlands Sixteenth Century Cartography: Mercator's Birth Pentecentennial', *Journal of Coastal Research* 32, no. 3 (2016), 670-85.
33. R.W. Cross, 'Dutch Cartographers of the Seventeenth Century', *Geographical Review* 6, no. 1 (1918), p. 68.

100 maps and charts.[34] The city at this time seemed awash with people involved in the map industry. In 1633, Henricus Hondius (Jodocus' son) made an even bigger splash when he published (in partnership with his brother-in-law, Johannes Janssonius) a two-volume atlas that included 239 maps.[35] Similar editions in German, Latin, Dutch, French and English followed and, towards the end of the decade, Hondius and Janssonius were publishing enlarged, three-volume atlases, bulging with an unprecedented 328 maps.

The Amsterdam atlases flooding the European market in this decade were luxury items. The expense in preparing and printing them was reflected in their price, and many ended up in the collections of wealthy bibliophiles, who cherished them as artistic artefacts maybe even more than as sources of geographical information. Behind the scenes, however, there was sometimes considerable disarray, as publishers sought to make a tidy profit from the mess of maps piled and filed in cartographers' studios. Old copperplate engravings were re-used (fair enough) but their use sometimes persisted for decades after it was known that the details on the maps were wrong.[36] Copper plates were traded among publishers, and it was not uncommon for an engraver's name to be replaced to give the impression that the map was a new creation. To muddy the waters even further, some engravers used the names of more famous cartographers in their family to give the works a greater sense of authority to map-buyers.[37] In a similar manner, signatures or authors' initials printed at the end of a passage of text in an atlas were sometimes altered, with methods ranging from simply crossing out the signature, to modifying it with a pen, stamping a new signature over it, or occasionally scratching it, but carefully enough to avoid creating a hole in the paper.[38] Coloured maps in these published collections were done by hand (using watercolours), with some of the more established cartographers employing their own colourists, although they would call on contractors during busy periods to ensure that their output met

34. P. van der Krogt, 'Selected Papers from the 16th International Conference on the History of Cartography: Amsterdam Atlas Production in the 1630s: A Bibliographer's Nightmare', *Imago Mundi* 48, no. 1 (1996), p. 151.
35. G. Mercator, *Atlas ou Representation du Monde Universel* (Amsterdam: Chez Henry Hondius, 1633).
36. Cross, 'Dutch Cartographers', p. 68.
37. Ibid.
38. Van der Krogt, 'Selected Papers', pp. 157-58.

demand. Unsurprisingly, haste affected the standard of colouring, as it did the consistency of the atlases. The maps were printed separately from the text of these publications, resulting in map sheets being shuffled and stockpiled as they awaited collation by book-binders. Thus, by the time an edition was published, usually no two copies were exactly alike.[39]

The vigorous business of cartography in Amsterdam was umbilically connected to the flourishing trading companies based in the city. They invested in mapmakers and demanded ever more precise charts to aid their mercantile activities. However, most cartographers were generally contracted by the trading companies rather than employed by them and, during any lulls in business, had to resort to publishing to make ends meet. Some, like Hondius and Janssonius, relied on the atlas market to bolster their revenues when company contracts were thin on the ground, whereas others, such as Gerritsz, could turn their skills to art engravings. By the late 1630s, the expansion of Dutch commerce internationally had led to the Dutch East India Company establishing its own cartographic workshops, offering the city's leading mapmakers more regular and certain employment. Cartography was now insinuated into the workings of these mercantile monoliths, with specialists in this field now part of the hives of those companies which were despatching expeditions out to the known ends of the earth.

39. M.P.R. van den Broecke, 'Unstable Editions of Ortelius' Atlas', *Map Collector* 70 (1995), 2-8; R.V. Tooley, *Maps and Mapmakers*, 6th edn (London: Batsford, 1978), p. 127.

Chapter 2

The VOC and Dutch Batavia

Van Linschoten's *Itenerario*

Maps did not just plot the course of Dutch commercial expansion. In the 1590s, they led the way, enabling ambitious merchants from the Netherlands to wrangle themselves into a global trading network that was dominated at the time by Europe's two imperial behemoths: Spain and Portugal. For centuries, the Low Countries had been a centre of the European trade in fishing, textiles and agriculture, but that continental prosperity was being eclipsed in the latter decades of the 1500s by intercontinental growth led by Spanish and Portuguese merchants, whose state-backed expeditions of conquest and consumption created lucrative international monopolies in the Americas and parts of Africa and Asia.

Slaves and spices were among the key goods enriching these two economic superpowers in this period, with clove, nutmeg, and mace in particular having been extracted by the Portuguese from around Indonesia since 1512.[1] For the Protestant Netherlands, however, muscling into this lucrative Catholic cartel using military force was not an option and so Dutch merchants resorted to a more effective weapon: efficiency. Spain, in particular, had invested heavily in a flotilla of warships to protect its trade routes and to prise open markets in the New World. The Dutch, in contrast, largely escaped the expense of

1. J. Setiawan and D. Kumalasari, 'The Struggle of Sultan Babullah in Expelling Portuguese from North Maluku', *HISTORIA: Jurnal Pendidik dan Peneliti Sejarah* 2, no. 1 (2019), 1-6.

a ponderous navy in favour of a less confrontational tactic. In one instance, for example, when attempting to make forays into the Caribbean, their merchants were faced with a Spanish naval blockade. Instead of retaliating, the Dutch ships simply sailed off in search of new markets elsewhere in the region where they could obtain the goods that they sought without Spanish interference.

This sort of prodding and probing by the Dutch in the realm of their competitors was possible because the geographic spread of the New World was so vast that even the most powerful European empires lacked the means to monopolise their trade in these regions entirely. It was a comparable situation with Portugal, which had a foothold in Asia but which, like the Spanish, was unable to defend every inch of the territories where it traded, and which similarly suffered from the corrosive inefficiencies inherent in monopolies. In Portugal's case, this resulted in its merchants struggling to meet growing demand for Asian goods in Europe, thus creating an opening for another trading nation to send ships to Asia to exploit this situation.[2] The Netherlands had the opportunity to capitalise on this unsatisfied demand, but not yet quite the means.

Other circumstances were converging at this time, however, to propel Dutch commercial expansion beyond Europe. The Spanish takeover of Antwerp in 1585 had forced many of that city's occupants, including most of its merchants, into exile and, by 1590, the majority of these traders had migrated to Amsterdam, bringing their substantial knowledge of international commerce and their business contacts with them.[3] This coincided with advances in Dutch shipbuilding and design, which, together with improvements in navigation techniques, enabled merchants in the Netherlands to reach more distant markets more frequently.[4] However, one final ingredient was needed for these developments to achieve the alchemy of empire-formation, and that was a set of maps. Without

---

2. J. Israel, *Dutch Primacy in World Trade, 1585-1740* (Oxford: Clarendon Press, 1989), p. 41.
3. Ibid., p. 42; O. Gelderblom and J. Jonker, 'Completing a Financial Revolution: The Finance of the Dutch East India Trade and the Rise of the Amsterdam Capital Market, 1595-1612', *The Journal of Economic History* 64, no. 3 (2004), 641-72.
4. V. Enthoven, *Zeeland en de opkomst van de Republiek: Handel en strijd in de Scheldedelta,* c. *1550-1621* (Leiden: Luctor et Victor, 1999), p. 248, in P.C. Emmer, 'The First Global War: The Dutch versus Iberia in Asia, Africa and the New World, 1590-1609', *Journal of Portuguese History* 1, no. 1 (2003), p. 6.

knowing the routes and destinations of their Portuguese competitors, the Dutch would remain as an economically marooned nation, confined to trading mainly with its terrestrial neighbours.

Of course, the Portuguese kept such information a closely guarded secret. The paranoia with which they had concealed this intelligence for a century was justified because it ensured that the monopoly they had on resource extraction remained secure. Yet, it was more blind luck than foresight that ensured that Portugal had kept confidential the maps and charts by which its economic empire functioned. After all, the messy mix of traders, officials and Church clerics – all with their own employees and entourages – had become bloated from the privileges deriving from the monopoly, as a result the barriers guarding the details of the country's shipping routes had become porous. Just how porous was exemplified in the case of the Dutch merchant Jan Huygen van Linschoten.

In 1579, at the age of sixteen, van Linschoten had moved from the Netherlands to Spain, to work with his older brother as a trader. A year later, he shifted to Portugal to take up a similar role and, through bumping into the right people at the right time, ended up being employed as the secretary to the Archbishop of Goa. While van Linschoten's livelihood came from the Catholic Church and the Portuguese state, his loyalties were rooted much more deeply in the Netherlands, where he had been born.[5] That loyalty never waned and, as he travelled extensively along the routes that Portugal's traders sailed to acquire cargoes of spices and other goods prized in Europe,[6] somewhere in the deep recesses of his imagination the idea of documenting his journeys for the benefit of his birth country germinated. Secretly, he jotted down details of trade routes, meteorological data, tides, port locations, the geography of Portuguese territories, details of the peoples and cultures of these places and, crucially, maps. There were maps for navigation, maps showing Portugal's colonial possessions, maps revealing the street layouts of trading posts, maps revealing ship movements, maps containing hundreds of place-names along the coasts of various countries, maps showing precise information on the river systems of South America, India and the Congo, maps displaying the topography of the islands of St Helena and

5. Van Linschoten was born in Haarlem, which was around one hour's horse ride from Amsterdam.
6. J. de Vries and A. van der Woude, *The First Modern Economy: Success, Failure, and Perseverance of the Dutch Economy, 1500-1815* (Cambridge: Cambridge University Press, 1997), p. 383.

Ascension, maps disclosing the position of the islands of Indonesia and maps of the world showing where all the other territories were located globally. In 1596, van Linschoten assembled the details from this unprecedentedly vast intelligence pillage and published them in a 460-page book, which he somewhat brazenly titled *Itinerario*.[7]

Itinerario was like an autopsy of the Portuguese Empire at this time, with all the entrails of navigation and trade locations spilling out in maps and texts for readers to pick through. It was not just Dutch readers, though, who were enthralled by the contents of this work. Van Linschoten's publisher, Cornelis Claesz, made an arrangement with a counterpart in London, John Wolfe, who produced an English version of the book in 1598, and other editions in French, Latin and German soon followed. Van Linschoten had mentioned in *Itinerario* that the Portuguese presence in India had become unstable and was afflicted by decadence.[8] He then achieved what probably no other person in history has done: to reorder European imperial power with the publication of a single book. In the few years that followed, Portugal's expansive trading empire, built on cartographical confidentiality, collapsed.

In the Netherlands, the sort of opportunism Van Linschoten had exhibited was regarded as an act of national virtue, enabling the country's merchants to unleash their entrepreneurial impulses in new territories and allowing its ruling classes to take what they were convinced was their rightful place in the pantheon of European maritime powers. Van Linschoten's book and maps were also an important part of the lineage leading to New Zealand's first map. They accelerated the gestation of the Dutch East India Company and eventually encouraged tentative ventures by the company's mariners into the South Pacific.

* * *

Almost immediately following *Itinerario*'s publication, Dutch forays into West Africa and Asia got under way. At first, they were barely viable financially; but they were immensely important in other ways. Technologically, they showed that Dutch navigators in Dutch ships were the equal of any others in Europe. Politically, these expeditions confirmed to everyone how rapidly and how completely Portugal's

7. J.H. van Linschoten, *Itinerario: Voyage ofte Schipvaert naer Oost ofte Portugaels Indien* (Amsterdam: Cornelis Claesz, 1596).
8. A. Saldanha, 'The Itineraries of Geography: Jan Huygen van Linschoten's *Itinerario* and Dutch Expeditions to the Indian Ocean, 1594-1602', *Annals of the Association of American Geographers* 101, no. 1 (January 2011), p. 149.

maritime prowess had flagged. Symbolically, residents in the Netherlands sensed that their country's time in the sun was dawning. At no previous point in its history had the prospects of prosperity seemed so attainable to this fledgling imperial power. However, the great Dutch commercial emporium that looked to be approaching on the horizon was still more a future hope than a present reality.

By 1600, the previously secret information that van Linschoten had appropriated from the Portuguese and then supplied freely to his countrymen had launched an armada of Dutch vessels off in search of fortune along the newly-revealed trade routes. Dozens of ships began to sail from the Netherlands each year for Africa, Asia and the Americas, encountering only sporadic Portuguese and Spanish resistance along the way. However, cost-cutting arising from competition between individual Dutch shipowners, combined with the European market being inundated with spices at certain times, led to fluctuating prices for the goods these traders brought back with them.[9] In the longer term, this was not the basis for a sustainable, profitable business, and so merchants and officials began to scramble for a solution that would achieve two seemingly opposite outcomes: an increase in the quantity of Dutch-imported goods into Europe; and the maintenance of the prices for those goods. It was an ambition that seemed to defy the basic economic laws of supply and demand, but one that was crucial for the Dutch if their budding trade-based empire was to have any chance of surviving and then succeeding.

There was no time for lengthy deliberation on this matter. The collapsing Portuguese monopoly was creating a global trade vacuum, and Dutch officials appreciated from the outset that, if they failed to capitalise promptly on these circumstances, some other nation would. Their resolution to this dilemma proved to be both practical and providential. On 20 March 1602, the Dutch government ordered that the numerous small trading firms in the country consolidate into a single trading company (originally known as the United East India Company, VOC). Although merchants had no choice in the matter, there was much to induce them to support the initiative. The authorities issued a charter granting its traders a monopoly trading right in Asia. This was a preemptive move, allowing Dutch merchants to feast on the carcass of the Portuguese monopoly in the knowledge that other potential predators

9. J. de Vries, 'Understanding Eurasian Trade in the Era of the Trading Companies', in M. Berg (ed.), *Goods from the East, 1600-1800: Trading Eurasia* (London: Palgrave Macmillan, 2015), pp. 7-39.

would be deterred by the prospect of retaliation from the Dutch government if the monopoly was interfered with.[10]

The VOC was unlike anything that had previously existed. From its inception, it was a beguiling *ménage* of business, politics and war. Primarily, it was a joint stock company (initially attracting over six million florins in capital from 1,800 investors, mainly merchants) with a multinational operation and was governed by a board of seventeen directors drawn from the largest shareholders.[11] Unsurprisingly, though, the state that had brought this vast entity into being was hardly likely to be a spectator looking on at the company's fate. Too much was at stake for such a hands-off approach. The government of the Netherlands ended up pouring subsidies into the company, in return for which it cultivated it as a military force. In the rush to praise the company's eventual financial triumphs, there has sometimes been a tendency to bury its earlier military dimension. The commitment to making the company a fighting force – in every sense – was enormously expensive and hobbled its profitability until the early 1620s.[12] However, the Dutch government's military capacity was greatly enhanced by the epiphytic relationship it had created with the company and, for decades, it served as the guarantee that the monopoly for its merchants in parts of Asia would remain unchallenged by other European powers.

With the company's rigging now in place, Dutch trading vessels set out for Asia on a sea of optimism. All the constituent parts – from investors, directors, officials and politicians, to boat-builders, merchants, sailors and soldiers – began working together like some single, expanding organism. One of the elements that made this growth possible was the emergence of a professional cartographic industry, centred in Amsterdam. Mapmakers were brought in, along with seasoned pilots, to instruct trainee navigators. Lessons were given on navigation, land-surveying, trigonometry and basic cartographical techniques.[13] In the

10. O. Gelderblom, A. de Jong and J. Jonker, 'The Formative Years of the Modern Corporation: The Dutch East India Company VOC, 1602-1623', *The Journal of Economic History* 73, no. 4 (2013), p. 1054.
11. F.S. Gaastra, 'The Dutch East India Company: A Reluctant Discoverer', *The Great Circle: Journal of the Australian Association for Maritime History* 19, no. 2 (1997), p. 110.
12. Emmer, 'The First Global War', p. 7.
13. B. Byloos, 'Nederlands vernuft in Spaanse dienst: Technologische bijdragen uit de Nederlanden voor het Spaanse Rijk, 1550-1700' (PhD thesis, Katholieke Universiteit, Leuven, 1986), pp. 33-35, referenced in Zandvliet, 'Mapping the Dutch World Overseas', p. 1435.

colonies, an apprenticeship system operated, in which novice mapmakers were trained by their superiors to provide at least the basic elements on which more detailed and professional maps could later be based.[14]

At breathtaking speed, the company grew to the point where it took on some of the traits of a nation. This was particularly noticeable from the 1620s on, when it established its overseas headquarters in Batavia. Its Governor-General in Asia, Jan Pieterszoon Coen, bullishly extended the company's territorial dominion and jurisdiction, and pumped profits back into the colony to strengthen its position in the region. Revenue was what really mattered, as Coen conceded to the company directors in 1622, suggesting to them that '[n]o great attention should be paid to the question of reputation or honour' for their own sake, on the basis that honour lay with the person 'who without doing unright or violence has the profit'.[15] This was not accomplished without some opposition, however. The dying beast of the Portuguese Empire was still able to muster enough strength on occasion to resist these Dutch incursions, which forced the company to engage in a protracted war with the Portuguese (and their Spanish ally). At the same time, another pretender was attracting attention in this part of Asia. The recently-formed English Honourable East India Company[16] was cautiously investigating what commercial prospects the region held. The Dutch were therefore forced into an elaborate diplomatic balancing act, with the two emerging maritime powers sometimes cooperating and sometimes in conflict with each other.[17]

The Importance of the Dutch Settlement at Batavia

As autumn deadened into winter each year, preparations in Amsterdam got under way for the departure of the Dutch East India Company's *Kerstvloot* ('Christmas fleet') to Batavia. These fleets were typically comprised of large, bulky cargo vessels, accompanied by a smattering of

---

14. Ibid., pp. 1434-35.
15. J.P. Coen, cited in N. Steensgaard, 'The Dutch East India Company as an Institutional Innovation', in M. Aymard (ed.), *Dutch Capitalism and World Capitalism* (Cambridge: Cambridge University Press, 1982), p. 255, in M. Bennett, 'Van Diemen, Tasman and the Dutch Reconnaissance', *Papers and Proceedings: Tasmanian Historical Research Association* 39, no. 2 (1992), p. 71.
16. S. Hejeebu, 'Contract Enforcement in the English East India Company', *The Journal of Economic History* 65, no. 2 (2005), p. 517.
17. V. Enthoven, S. Murdoch and E. Williamson (eds), *The Navigator: The Log of John Anderson, VOC Pilot-Major, 1640-1643* (Leiden: Brill, 2010), p. 1.

smaller transport ships and a warship to fend off pirates, the Portuguese, the Spanish or the English (the distinction between the vessels of these nations and pirates was not always clear).[18] Winter was a challenging time to depart, with sailors having to contend with the stormy and bitterly cold North Sea as they sailed from the Netherlands. However, by way of compensation, it also meant that the fleets would encounter more favourable weather on the final leg of their journey to Batavia. Furthermore, it was also generally easier to recruit sailors in Europe over winter, when work locally was thinner on the ground. As additional vessels bolstered the company's fleet by the 1610s, the departure schedules were expanded. There was the *Kermisvloot* ('Fair fleet'), which left Amsterdam around the time of the city's September fair and was generally smaller in size (usually carrying mail and domestic supplies rather than trade goods). Then, in spring, the larger *Paasvloot* ('Easter fleet') set off for Asia, completing the company's annual trade cycle.[19]

The pride of these fleets, or their workhorses, depending on how you looked at it, were the East Indiamen. These were 700-ton, square-rigged, sailing ships that were designed, built and operated by the company and that were capable of carrying up to 300 people, but which were ill-suited to sailing windward and also notoriously difficult to navigate in storms and in confined waters.[20] To a degree, the physical limitations of these vessels were offset by the excellent cartographical information that was lavished on their captains. Lavish is an appropriate term because these were works of exquisite composition accompanied by suitably modest, although still expert, ornamentation. Amsterdam's cartographers had been kept busy throughout this era, stitching together any intelligence about the company's trading routes that was sent to them and modifying their maps and charts accordingly. Through the trial and error of company mariners, and the observations they meticulously recorded, Amsterdam's mapmakers were producing works of great accuracy,

18. T. Andrade, 'The Company's Chinese Pirates: How the Dutch East India Company Tried to Lead a Coalition of Pirates to War against China, 1621-1662', *Journal of World History* 15, no. 4 (2004), 415-44.
19. F.S. Gaastra, *Geschiedenis van de VOC: Opkomst, bloei en ondergang* (Amsterdam: Amsterdam University Press, 2016), pp. 116-17, in C.E. Ariese, 'A twisted truth – the VOC ship *Batavia*: Comparing history & archaeology' (BA dissertation, Leiden University, 2010), p. 10.
20. J. Gawronski, 'East Indiaman *Amsterdam* Research 1984-1986', *Antiquity* 64, no. 243 (1990), 363-75.

which made the voyages of the fleets more secure. These charts included more than just a layout of locations. Navigation lines radiated in all directions, and details on winds, tides and hazards were included and, like everything else, subject to constant updating as new information about the passages to and from Asia made its way back to the company's headquarters.

Towards the end of the year, as preparations gathered pace for the departure of the *Kerstvloot*, activity at the port at Amsterdam would become even more frantic. Hunched workers lugged sails and ropes along the docks, while passengers, sailors, and merchants bustled their way through, trying to find particular ships, checking on inventories and making last-minute arrangements. Everyone was battling the cold. Although light flurries of snow might waft through, making surfaces even more slippery, rain was the greater enemy. It saturated stores and soaked though the coats and capes of those loading the ships, making their movements even more sluggish. Often, mists chilled by air from the Arctic billowed in. Visibility would then be reduced to the point where only the forest of ships' masts would be visible, making departure too risky. Still, though, preparations continued. Crates containing supplies were shoved and wheeled onto decks and into hulls, cattle and poultry were corralled into spare spaces, scores of last-minute preparations were made, inspectors and officials completed their paperwork and goodbyes were said. The arrangements were so considerable because the risks were so great. Once on the open ocean, a ship became a cocoon, having to contain every necessity of life for its crew for several months at a time.

When the head of the fleet was satisfied that everything and everyone was ready for departure, and when the weather obliged, the ships were finally able to be released from their moorings and leave. To the sounds of creaking rigging, sails being whipped by the breeze and sailors hoarsely yelling in every direction, the fleet embarked on the shortest leg of its voyage – just 100 kilometres north to Texel. This floorboard-flat island was used as a water supply by Dutch sailors setting out on long voyages. The water in the island's wells was rich in iron, which meant that it would take longer to spoil than water obtained on the mainland.[21] Texel was also a good location for the fleet to pause and gather the latest intelligence on shipping activity in the vicinity. If the English Channel was judged to be fairly free of risk from English pirates, the company

21. J. Haarhoff, 'Water and Beverages on DEIC Ships between the Netherlands and the Cape: 1602-1795', *Historia* 52, no. 1 (2007), 127-54.

ships would then sail south, sometimes joining vessels affiliated with the company's offices in Delft, Rotterdam and Zeeland. However, if the Netherlands' fluctuating relations with England were at one of their periodic low points, the company flotilla would head northwest, arching around Scotland before making its way south.[22]

One of the more important navigational breakthroughs for these fleets occurred in 1610, when Dutch maps were amended to depict a new course between the Netherlands and Batavia. From that time, the nation's ships would hug the West African coast as far as modern-day Gambia, and then plunge in a south-westerly direction for the next 4,500 kilometres into the South Atlantic Ocean. From there, the fleets would swerve directly east towards the Cape of Good Hope, where they would dock to be refitted and re-supplied for the next leg of the voyage. This final stage involved a 7,500-kilometre route, with the surging winds of the Roaring Forties driving the ships swiftly eastward. Then, just before reaching the eastern coast of Australia, the fleet would again abruptly change direction – like a flock of birds – this time turning north for the final leg (around 3,500 kilometres) to Batavia.

This course was pioneered by Hendrik Brouwer, from which it derived its popular name: the Brouwer Route. It made the journey faster and reduced the risk of encountering violent seasonal storms. Shorter travelling times also meant that provisions were less likely to go bad and the health of the crew would be better safeguarded as a result. The Brouwer Route was so effective that, in 1616, the company mandated that it be the official course for all its ships sailing to the region.[23] However, there was one persistently frustrating impediment to this otherwise expertly-mapped passage: humanity had yet to work out a way of measuring longitude.[24] As a result, captains would have to estimate how far east they had sailed from the Cape of Good Hope before turning north. If the distance was underestimated, they might end up on the arid shores of Australia's west coast. If it was overestimated, they would

22. E. Hartkamp-Jonxis, *Sits: Oost-West Relaties in Textiel* (Zwolle: Waanders, 1987), pp. 1-6.
23. J.R. Bruijn, 'Between Batavia and the Cape: Shipping Patterns of the Dutch East India Company', *Journal of Southeast Asian Studies* 11, no. 2 (1980), p. 256; A.W. Beasley, 'The First Amputation in Australia', *Seventeenth Century* 17, no. 1 (2002), p. 72.
24. H. Drake-Brockman, *Voyage to Disaster: The Life of Francisco Pelsaert Covering His Indian Report to the Dutch East India Company and the Wreck of the Ship* Batavia *in 1629* (London: Angus & Robertson, 1982), p. 41.

arrive in India instead of Indonesia. Tackling this challenge relied on the experience of the ships' captains, who daily had to determine how far their vessels had sailed and then, when they estimated that the desired distance had been covered, change their direction to point to Batavia. Copies of instructions for traversing this route were handed to every captain and mate before departing the Netherlands, although it was only as late as 1652 that the first printed versions of the map were issued.[25] This was because maintaining a dependence on hand-drawn charts lessened the possibility of the documents entering into the hands of competitors[26] (the very thing that had led to the Dutch superseding Portuguese commercial dominance in Asia).

By the 1630s, the increase in the number of journeys made by the company's vessels – which went hand-in-glove with improved maps – had led to the route from the Netherlands to Batavia being nicknamed the *wagenspoor* ('cart track')[27] but more for the frequency of its use rather than any certainty it offered. Familiarity did not breed security in this case: and it was not just the dangers that came from the weather or navigational miscalculation. The threats the fleets faced were often primarily onboard ones. The food would usually turn rancid after around a month at sea, and crew and passengers sometimes suffered disease, which found an easy roaming ground among the ships' cramped quarters and poorly-nourished occupants. The greatest health threat that plagued passengers, though, was scurvy, caused by a deficiency of vitamin C. It could lead to anaemia, exhaustion, aching and swollen limbs, spontaneous bleeding, ulcerated gums, a loss of teeth and, occasionally, death.[28] Still, though, there was always a willing queue of men eager to serve on the company's vessels, mainly because the pay offered was better than that which could generally be obtained on the mainland and because promotion prospects were good (even if only due to the fact that death-rates on the journeys were so high that, over time, survivors found their competition often severely reduced in number).[29]

One telling sign of how commerce reigned as king on these voyages was in the hierarchy that prevailed in the fleets. There was the crew, then

25. Bruijn, 'Between Batavia and the Cape', p. 257.
26. Schilder, 'Organization and Evolution', *Imago Mundi* 28 (1976), p. 62.
27. Enthoven, Murdoch and Williamson (eds), *The Navigator*, pp. 31-57.
28. J.V. Hirschmann and J.R. Gregory, 'Adult Scurvy', *Journal of the American Academy of Dermatology* 41, no. 6 (1999), 895-906.
29. Gaastra, *Geschiedenis van de VOC*, pp. 91-100, in Ariese, 'A twisted truth', p. 11.

assorted extras (cooks, barbers, carpenters) and, a level above them, doctors. Further up the ladder were the navigators and captains but, looking down on all, were the senior merchants who travelled with their cargoes. These 'upper merchants', as they were called, were conferred with the rank of 'commandeur' and were partly in charge of the fleets – only partly, though, because a council of merchants and captains also had a significant role decision-making on these voyages.[30] When it came to navigation, though, they were obliged to act in partnership with the captains of the ships in the fleet, on whose nautical experience the commercial success of the expeditions depended. The result was an opaque command structure, but one that nonetheless worked effectively most of the time.

If the daily measurements of distance travelled were sufficiently accurate, and if the commanders and captains of the fleets had adhered to their charts and detailed navigational instructions, then in the final days of their voyage, the landmass they would see on the horizon would be West Java. Using the island of Krakatoa as a guide, these vessels would navigate through the Sunda Strait, past the company watch post at Anjer-Kidu, and then sail the final 130 kilometres to Batavia.

Maps, as much as any other facet of the VOC's operations, safeguarded this vast enterprise and so, from the beginning of the seventeenth century, efforts were made to standardise surveying methods and mapmaking to ensure that this colossal commercial venture was not undermined in any way by inconsistent cartographical information. One example of this standardisation was the *Rijnlandse roede* ('Rhineland rod'). This was a surveying unit of 3.767 metres.[31] Of these rods 5.5 metres amounted to a chain, which became the set measure for public works and land administration (although in instances where accuracy was less important, paces remained the favoured unit). By 1642, the rules governing these standardised measurements had been metaphorically carved in stone in the form of the Batavia law codes (*Bataviasche Statuten*).[32] These rules included responsibilities for surveyors and administrators to adhere to these standardised measures, and for details of the mapping of any

30. Ariese, 'A twisted truth', pp. 11-12.
31. A. Gomperts, A. Haag and P. Carey, 'Mapping Majapahit: Wardenaar's Archaeological Survey at Trowulan in 1815', *Indonesia* 93 (2012), p. 180.
32. J. van Kan, 'De Bataviasche statuten en de buitencomptoiren', *Bijdragen tot de Taal-, Land- en Volkenkunde van Nederlandsch-Indië* 100 (1941), 255-82, referenced in Zandvliet, 'Mapping the Dutch World Overseas', p. 1436.

territory to be confirmed by officials in Batavia and in the Netherlands before any construction work commenced.[33] Thus, Amsterdam and Batavia were bound by both a sailing route and, increasingly, shared surveying and mapping conventions. Standardisation in terrestrial maps soon spread to their nautical counterparts and, all the time, the accretions of cartographical knowledge continued to accumulate, literally outlining Dutch dominance in various parts of the world outside Europe.

Company cartographers were given access to all the logs, journals, notes and charts that were used by each fleet, so that corrections could be made to their existing maps, with the improved versions then passed on to the next fleet. It was a system of constant refinement, but one which was also subject to external monitoring. The company directors in Amsterdam had to approve all but the most minor emendations to its maps. Not only did this maintain consistency and quality, but it also ensured that the directors were kept abreast of each new contour of their growing international enterprise. Once completed, the newly revised charts and maps were kept in a secure room in the company's headquarters in Amsterdam, with the cartographers having to take an oath (before the mayor of the city) that they would not disclose any details of their work to anyone outside the company.[34]

By the end of the 1610s, though, it was reaching the point where the piles of cartographical information held by the company were becoming almost unmanageable. The relentless drive by the company's directors to accumulate more nautical and geographical details about their trading empire made sense when it came to bringing more efficiency to their transport routes. However, the result was that their map collection was bursting at the seams. In response to a query, company archivists would have to leaf their way through this forbidding jungle of papers of all sizes – 20 years of cartographical material, which was lodged in shelves, filed in drawers and cupboards, and stacked on tables, without any coherent archiving or indexing system. Thus, in 1620, the directors wrote to Hessel Gerritsz, instructing him to consolidate some of this information into a new publication so that it would be more accessible to officials and mariners. The governor of Batavia was ordered to send copies of all the drawings, maps and surveys he had amassed to assist Gerritsz with this enormous undertaking, which aimed to condense

33. Zandvliet, 'Mapping the Dutch World Overseas', pp. 1435-36.
34. Schilder, 'Organization and Evolution', p. 62.

all the accumulated knowledge of the East Indies into one volume. In June 1622, Gerritsz proudly informed the company directors that he had completed his task and that the book was ready for use.[35] However, neither the book, nor the original maps it contained, have survived,[36] although copies of some of the charts did escape oblivion, including one of the Pacific, which shows the ocean, flanked by South America on the right, and the East Indies on the left, with the Solomon Islands and New Guinea as the southernmost territories depicted in that latitude.[37]

Until this time, Dutch captains sailing through the East Indies still relied on older Portuguese charts to assist them. These tended to be less accurate and reliable, jeopardising the precious cargoes on which the company's existence depended. From 1618, Dutch-produced charts began to displace those of their Portuguese predecessors and, two years later, a hydrographic office was established in Batavia, to accelerate the process of updating maps as new cartographical information came to light.[38]

The need to make order out of chaos – which had been responsible for the company commissioning Gerritsz in 1620 to make a volume that collated and made accessible bundles of papers on Dutch navigation – was replicated in Batavia in the following decade. By 1632, there were already senior mapmakers working in this Dutch East India Company capital in the Southern Hemisphere, and most of them had close connections with their counterparts in Amsterdam (one of those cartographers who commenced work at this time in Asia was one of Gerritsz's sons). As Batavia's centrality to the Dutch East India Company's trade grew, so

---

35. Algemeen Rijksarchief, Koloniaale Afdeling (State Archives, Colonial Department The Hague), 452, Missive Heren XVII to J.Pz. Coen, 9 September 1620; 4653 D 207 b. Missive by Hessel Gerritsz to Adriaen van Santvoort, 23 June 1622, in Schilder, 'Organization and Evolution', p. 77; Keuning, 'Hessel Gerritsz', p. 56.
36. Schilder, 'Organization', pp. 65, 68-69.
37. H. Gerritsz, *Extrait de la carte du Pacifique d'Hessel Gerritsz (1622), montrant les 'îles Salomon' ainsi que 5 îles découvertes par Jacob Le Maire et Willem Schouten en 1616*, Bibliothèque nationale de France, Carte conservée au département des Cartes et plans, cote GE SH ARCH-30 (RES); M. Woods, 'For the Dutch Republic, the Great Pacific', in National Library of Australia, *Mapping our World: Terra Incognita to Australia* (Canberra: National Library of Australia, 2013), pp. 111-13.
38. Schilder, 'Organization and Evolution', p. 72; Zandvliet, 'Mapping the Dutch World Overseas', p. 1443.

too did the demand for maps of this part of the world. Around 1640, a new role – Supervisor of Navigation – was established in the colony, and the Hydrographic Office, which was built around the Supervisor of Navigation, augmented its cartographical work by committing itself to a new field of activity: exploration.[39]

* * *

Just the mention of the Batavians was enough to send seventeenth-century Netherlanders into an indulgent moment of reflected glory over their founding myth.[40] This ancient Germanic tribe, which occupied the area south of what later became Rotterdam, was seen to have embodied all the national virtues to which the late-Renaissance Dutch aspired. Thus, the symbolism of giving the name Batavia to the settlement of Jakarta, on the island of Java, could not have been greater. In doing so, the Dutch East India Company was imprinting not just its commercial but also its imperial and cultural ambitions on the marshy flatlands of this diminutive port settlement, in the earnest hope that it would eventually become the centre of its hydraulic empire.

It was usually in July each year that the Christmas fleet limped into the harbour at Batavia after the titanic final 10,500-kilometre leg of its voyage. By the time that the battered vessels docked, the crews were often afflicted with sicknesses, injuries and gnawing hunger. In their weakened state, they disembarked, at which point some would succumb to the temptation to gorge on the abundance of tropical fruit being sold by street vendors who crowded the port as the fleet arrived. The result of a sudden heavy dose of vitamin C to those who had practically gone for months without it was violent sickness[41] – and this was just the first of many adjustments that Dutch arrivals faced in what was for them an otherworldly environment.

Yet, despite the malnutrition, the illnesses, the arduous nature of these voyages of attrition and an unfamiliarity with the destination, there was still a ready supply of men willing to work on the fleets sailing to Batavia. As early as 1610, the total number of crew on company vessels was around 33,000, which represented six per cent of all Dutch

39. Zandvliet, 'Mapping the Dutch World Overseas', p. 1443.
40. J. Leerssen, *National Thought in Europe: A Cultural History* (Amsterdam: Amsterdam University Press, 2006), pp. 39ff., in Kehoe, 'Dutch Batavia', p. 5.
41. K.J. Carpenter, *The History of Scurvy and Vitamin C* (Cambridge: Cambridge University Press, 1988), p. 59.

men over the age of fifteen.[42] Few other commercial entities in Europe at that time (or, indeed, since) could claim that proportion of a nation's workforce.

Almost from its inception, Batavia was more than a trading post. Socially, company directors envisaged it emerging as a utopian miniature of Dutch society – epitomising all its values and eschewing all its vices. Of course, the Dutch citizens were a minority on the island, outnumbered by Chinese, Indian and indigenous residents, as well as slaves who had been brought to Java to supplement local labour. However, population size was not the criterion by which dominance was determined. The Dutch imposed systems of power – ranging from street layouts to administrative architecture, the built environment and even the forms of clothing that were permitted.[43] (The result of the latter was a lingering tension between the urge to be ostentatious – a natural manifestation from the wealth being generated in Batavia – and the strictures of modesty that were idealised in Dutch Calvinism.[44])

The means by which the minority Dutch asserted their supremacy in Batavia was sometimes a source of social strain. Outnumbered by other cultures and ethnicities, and in a location vastly foreign to that of north-western Europe, company officials cultivated Dutch dominance by every means within their reach. Populations were separated geographically by their ethnicity, with town precincts set up for different groups – many conveniently divided by canals which ran through the settlement (surely an aquatic reminder of Amsterdam). As this was to be the capital of a commerce-state, the scale and grandeur of buildings were intended to reflect the relative status of their occupants, with the more substantial houses overburdened with various Dutch architectural motifs for good measure.[45] Yet, for all the attempts at social, cultural and ethnic segregation, integration of the various groups in Batavia could not be prevented. A painting of the settlement from the middle of the seventeenth century by the Amsterdam artist Andries Beeckman revealed a more cosmopolitan populace mixing with each other, with the most obvious sign of division being the incongruity of

42. Emmer, 'The First Global War', p. 12.
43. J.G. Taylor, *The Social World of Batavia: European and Eurasian in Dutch Asia* (Wisconsin: University of Wisconsin Press, 2004), pp. 68-69.
44. A.M. Kettering, 'Gentlemen in Satin: Masculine Ideals in Later Seventeenth-Century Dutch Portraiture', *Art Journal* 56, no. 2 (1997), 41-47.
45. Kehoe, 'Dutch Batavia', p. 16.

austere, Dutch-styled buildings rising from a palm-peppered tropical landscape.[46] It was into this slightly surreal, social, political and urban experiment that the Dutch mariner Abel Janszoon Tasman arrived for the first time around 1633.

---

46. A. Beeckman, *The Castle of Batavia, seen from Kali Besar West*, Rijksmuseum, Object No. SK-A-19.

Chapter 3

Abel Janszoon Tasman

Sailor, Captain, Explorer

A glimpse of Tasman's personality in paint peers through the gloomy varnish in Jacob Gerritsz Cuyp's portrait of the navigator and his family, which was completed sometime around 1637.[1] Tasman's watery blue eyes stare thoughtfully into the middle distance while he gestures towards a globe, clasping a drawing compass in one hand as he does so – and, just in case the viewer misses these clear cartographic cues, hanging on the wall in the background is an astrolabe, which was used by sailors to measure latitude, and that had come to symbolise the science of seafaring in this era. This is not a depiction of a rapacious coloniser, but of a navigational specialist. Curiosity, not dominion, is the motif in this portrait. For men such as Tasman, plotting routes and sailing across oceans were ends in themselves rather than a means to any imperial aggrandisement. They were paid functionaries, employed by the Dutch East India Company principally as sailors and navigators, ensuring the regular flow of trade shipments between ports.

Six years before Cuyp painted this portrait, Tasman was a mere *vaerentgesel* – a sailor, with probably no navigational training. However, since then, he had served on company ships in South East Asia and had accumulated considerable knowledge about sailing and navigating in the process. By the time Tasman entered Batavia's nautical orbit, in 1634, company officials were keen on probing

1. J. G. Cuyp, *Portrait of Abel Tasman, his wife and daughter* (*c.* 1637), National Library of Australia, call number PIC T267 NK3.

Portrait of Abel Tasman, his wife and daughter (Jacob Gerritsz Cuyp, 1637)
National Library of Australia

beyond their existing trade routes to see what opportunities might be discovered in some of the more unfamiliar areas in the region – the sort of enterprise that required captains shifting from being map-readers to mapmakers. On 18 February that year, Tasman, who had now ascended to the position of first mate on the trading ship *Weesp*,[2] was despatched from Batavia on one such exploratory journey. The ship's destination was Ambon Island, around 2,200 kilometres east of Batavia – a location the Dutch had first visited in 1599, and where they were now asserting a firmer presence.[3] By late May 1634, Tasman had risen to the rank of captain and was put in command of the *Mocha*, which was despatched that month to map safer routes for Dutch ships

2. K. Schuldt, 'Abel Janszoon Tasman', *Deutsches Schiffahrtsarchiv* 8 (1985), 117-46.
3. P.V. Lape, 'Political Dynamics and Religious Change in the Late Pre-colonial Banda Islands, Eastern Indonesia', *World Archaeology* 32, no. 1 (2000), p. 149.

in the region.[4] It was a small and rudimentary undertaking and one which he accomplished to the satisfaction of his superiors in Batavia.[5] In August 1635, he was promoted to the rank of commander, heading a small fleet of vessels which served to protect company trade in the region,[6] and on 30 December 1636, the now much more weary and weather-worn Tasman boarded the *Banda* for the gruelling voyage back to Amsterdam, arriving on 1 August 1637. Eight months later, having settled his affairs in the Netherlands, Tasman sailed back to Batavia (this time, accompanied by his wife), reaching the port on 11 October 1638. Within weeks, he was again captaining a ship, this time headed for Ambon Island, where he took delivery of a cargo of cloves before returning to Batavia.

While Tasman was completing this relatively mundane voyage, the local company directors were already concocting his next mission – one that would be almost entirely exploratory in nature. Batavia's blossoming bureaucracy was burdening the company with mounting costs, for which increased trade was seen as the principal panacea. To that end, a plan was devised for two vessels (one captained by Tasman) to sail roughly 4,000 kilometres north in the direction of Japan, with instructions to chart 'unknown lands, shores, and shoals or shallows' along the way and to ensure that all the intelligence gathered was kept strictly confidential.[7] A great deal of navigational and geographical information was harvested as the two ships tacked their way through the seas of South East Asia.

While the Dutch predictably painted their activities in this part of the world in favourable hues, some of those at the receiving end of their ventures were sometimes less approving. A Japanese assessment made at this time offers a blunt corrective, describing the Dutch as '[g]reedy and cunning, good judges of precious commodities' and saying that they 'fight with skill to obtain the greatest possible profit'. This was followed by a stark warning: '[w]hoever meets them on the high seas will assuredly be looted by them'.[8] The Dutch certainly enforced

4. A. Hoving and C. Emke, *The Ships of Abel Tasman* (Hilversum: Uitgeverij Verloren, 2000), p. 14.
5. J.E. Heeres and C.H. Coote (eds), *Abel Janszoon Tasman's Journal* (Los Angeles: Kovach, 1965), p. 10.
6. J.B. Walker, *Abel Janszoon Tasman: His Life and Voyages: Read before the Royal Society of Tasmania, 25th November, 1895* (Hobart: Government Printer, 1896), pp. 9-10.
7. Heeres and Coote, *Abel Janszoon Tasman's Journal*, p. 25.
8. Cited in P. Zumthor, *Daily Life in Rembrandt's Holland* (London: Weidenfeld & Nicolson, 1962), p. 297, in Bennett, 'Van Dieman, Tasman and the Dutch Reconnaissance', p. 71.

their monopoly trading routes 'ruthlessly' and their patrolling vessels, including those captained by Tasman, were prepared to sink any ships carrying illicit cargoes.[9]

* * *

Tasman returned to Batavia on 19 February 1640 and delivered his journals to the company's office. The maps were far from comprehensive, but the company was not looking for anything definitive anyway. Rather, the incremental accumulation of intelligence was what mattered and, by this time, for the board of directors in Batavia, acquiring details on little-known territories or new sailing routes was becoming a much greater priority for the company.

After four months in the cramped confines of Batavia, Tasman was once again about to sail into the yawning expanse of the South China Sea. On 14 June 1640, at the helm of the *Oostcappel*, he led a fleet of four vessels northwards to Taiwan, Japan (where he remained for four months) and then Cambodia, before returning to Batavia on 11 April 1641. Four weeks later, he was again captaining the *Oostcappel* to Cambodia and Japan, straggling back into Batavia on 20 December after being battered on the final stretch home by a vicious storm that blasted through the flotilla, fatally damaging two of the ships, which along with their crews ended up entombed in the ocean. Tasman's vessel just survived, but he was given little time to recuperate. Almost immediately, he was sent back on another voyage – this time a trading trip to nearby Sumatra, which concluded on 20 June 1642. For the following seven weeks, Tasman adjusted to landbound life while waiting for orders for his next command. His reputation as a competent and reliable captain was now well established and justified and so, as the Dutch East India Company assembled its plan for an exploratory voyage to as-yet-uncharted areas to the west and south, Tasman emerged as an obvious candidate to lead this expedition. It probably helped that he also privately coveted the role, knowing that it would boost both his prestige and his pay.

* * *

The equatorial heat and humidity were unremitting in Batavia. On 13 August 1642, as the sun lifted above the morning haze hanging over the mangrove forests and mudflats to the east of the settlement, even the tide seemed to lap lazily against the ships moored in the harbour. Slowly,

9. Ibid., p. 72; A. Sharp, *The Voyages of Abel Janszoon Tasman* (Oxford: Clarendon Press, 1968), pp. 4-5.

though, the town began to bustle into life as its residents commenced their daily activities. There was certainly nothing listless in the company's office that morning, though. Despite the overbearing heat, Dutch officials were busy putting the finishing touches to a meandering set of instructions that would form the basis of Tasman's exploration of parts of the world that no European had previously visited, and that would produce a new body of maps for the company.

The instructions arose from a resolution passed by the Governor-General and councillors of Batavia twelve days earlier, which in turn had originated in the minds of the company's directors back in Amsterdam. The general intention was to explore 'the partly known and still unexplored South- and East-land', in order to discover the most 'convenient routes to well-known opulent markets, in such fashion that the same might in due time be used for the improvement and increase of the Company's general prosperity'. The seed for the idea of this expedition had been planted in 1638 by Anthoonij van Diemen (Governor-General of the Dutch East Indies in Batavia, 1636-45), who had urged the directors to give 'further attention to the discovery of the South [Land] and the gold-bearing island, which would be of great use to the Company, in order in time to get over the heavy burdens, and come into the real enjoyment of the profits of the East India trade'.[10]

This would be an expedition involving only two vessels: the *Heemskerck*, a 60-ton 'yacht'; and the 100-ton *Zeehaen*[11] – placed under the overall command of Tasman, who was known by officials in Batavia as being 'strongly inclined to this discovery'. Frans Visscher would captain one of the ships and Tasman the other and, crucially, they would be accompanied by the merchant Isaack Gilsemans, who was described as 'sufficiently versed in navigation and the drawing-up of land-surveyings'.[12]

There was a particular line of logic threaded through this mission. Company leaders had looked back to the discoveries of Christopher Columbus and Amerigo Vespucci in the Americas, and Vasco de Gama in Africa and the East Indies, and reasoned that a similar feat could be replicated in the Pacific. There was thus a sense of history that was invigorating the urge to explore, in addition to the prospect of profits.[13]

---

10. Van Diemen, cited in Bennett, 'Van Dieman, Tasman and the Dutch Reconnaissance', p. 75.
11. R. McNab, *From Tasman to Marsden: A History of Northern New Zealand from 1642 to 1818* (Dunedin: J. Wilkie & Co., 1914), p. 2.
12. Heeres and Coote, *Abel Janszoon Tasman's Journal*, pp. 129-30.
13. Bennett, 'Van Dieman, Tasman and the Dutch Reconnaissance', p. 68.

It also seemed to make some sort of intuitive sense that a huge territory lay somewhere to the south. The mythical 'Province of Beach' – a purportedly large and lucrative land-mass – had been referred to in maps for two centuries before Tasman's expedition.[14] It was a speculative location, but one that the company now had the means to investigate. Tasman was furnished with the records of 'certain experienced pilots', who had previously plotted some of the routes south of Batavia, but which were of little practical use when penetrating areas of the ocean that were still completely uncharted.[15] Detail would be paramount, as the company made explicit to Tasman:

> All the lands, islands, points, turnings, inlets, bays, rivers, shoals, banks, sands, cliffs, rocks etc., which you may meet with and pass, you will duly map out and describe, and also have proper drawings made of their appearance and shape, for which purpose we have ordered an able draughtsman to join your expedition; you will likewise carefully note in what latitude they are situated; how the coasts, islands, capes, headlands or points, bays and rivers bear from each other and by what distances they are separated; what conspicuous landmarks such as mountains, hills, trees or buildings, by which they may be recognised, are visible on them; likewise what depths and shallows, sunken rocks, projecting shoals and reefs are situated about and near the points; how and by what marks these may most conveniently be avoided; item whether the grounds or bottoms are hard, rugged, soft, level, sloping or steep; whether one should come on sounding, or not; by what land- and seamarks the best anchoring-grounds in road-steads and bays may be known; the bearings of the inlets, creeks and rivers, and how these may best be made and entered; what winds blow in these regions; the direction of the currents; whether the tides are regulated by the moon or by the winds; what changes of monsoons, rains and dry weather you observe; furthermore diligently observing and

14. J. Huttich and S. Grynaeus, *Novus orbis regionum ac insularum veteribus incognitarum* (Basel: Hervagius, 1532), in T. Suarez, *Early Mapping of Southeast Asia* (Hong Kong: Periplus, 1999), p. 160; L.C. Wroth, *The Early Cartography of the Pacific* (New York: Bibliographical Society of America, 1944), pp. 5-8, in Sharp, *The Voyages of Abel Janszoon Tasman*, p. 24.
15. Heeres and Coote, *Abel Janszoon Tasman's Journal*, p. 132.

> noting whatever requires the careful attention of experienced steersmen, and may in future be helpful to others who shall navigate to the countries discovered.[16]

As if the risks of sailing through unknown parts of the world's largest ocean were not daunting enough for these explorers to contend with, the company accompanied its mapping instructions with a warning about the need for 'extreme caution' due to the likelihood of 'fierce savages' in some of the territories that they might come across. The experience of other European powers had taught the Dutch 'that barbarian men are nowise to be trusted, because they commonly think that the foreigners who so unexpectedly appear before them, have come only to seize their land'. Unlike their predecessors, however, the Dutch were aware of the reason for such indigenous suspicion, which they attributed to the 'many instances of treacherous slaughter' perpetrated by the Spanish and Portuguese in the Americas.[17]

The day after Tasman received his instructions, the voyage of the *Heemskerck* and *Zeehaen* got under way, despite the company's preparations for the expedition being 'hopelessly unsatisfactory'.[18] Perhaps the wisest decision was in the timing of the venture. By departing in August, the ships would be conducting the bulk of their exploratory work during the southern-hemisphere summer, thus offering longer daylight hours for those on the vessels to make and record their observations.[19] With favourable conditions and strong, dependable winds filling their sails, the two ships were whisked westward, sighting the white, foam-rimmed coral reefs of Mauritius on 5 September. While the crews used their time in port on the island to repair their ships (both of which were in extremely poor condition),[20] the Dutch Governor of Mauritius, Adriaan van der Stel, handed Tasman a collection of maps and charts. Some were useful (such as those detailing the Solomon Islands) and some of practically no value to the expedition (those charting the Straits of Magellan, for example). However, they were all pieces that contributed to the mosaic of cartographical knowledge to which Tasman had been instructed to add. From this point onwards, the route Tasman planned

16. Ibid., pp. 132-33.
17. Ibid., p. 134.
18. Ibid., p. 111.
19. Bennett, 'Van Dieman, Tasman and the Dutch Reconnaissance', p. 68.
20. Walker, *Abel Janszoon Tasman: His Life and Voyages*, p. 31.

to take would be into 'nameless waters' in an endeavour to 'enlarge the world' as Europeans knew it.[21]

On 8 October, as the ships prepared to depart – brimming with supplies and expectations – Tasman unrolled the map that laid out the all the known landmasses in this maritime empire and again let his eyes stray over the immense uncharted expanses of ocean beyond them. It was a forbidding cartographical blank space, stretching roughly 20,000 kilometres from the coast of Africa in the west to the shores of Chile in the east. This was the known as 'The Great Chart of the South Sea' and was based on Gerritsz's 1622 map of the region.[22] Another source that Tasman had access to at this time was Antonio de Herrera's *De Nieuwe Werelt*, published the same year, which contained maps of parts of the western Pacific and the Solomon Islands,[23] which was consulted along with a globe on board the *Heemskerck* (although the precise version of the globe used on this voyage remains unknown).[24]

From Mauritius, Tasman sailed southeast, reaching as far south as 49 degrees longitude on 6 November. There, the commanders of the ships decided to head more or less directly east and continued for another eighteen days until in the late afternoon of 24 November, they sighted land 'not as yet known to any European nation'. It was bestowed with the name Anthoonij van Diemenslandt, in honour of the Batavian Governor-General. This was later contracted and anglicised as Van Diemen's Land (which in 1856 became 'Tasmania' – ironically, a name given to the island by British settlers).[25] Two mountains towering over the landscape also succumbed to the European habit of applying names to almost every topographical feature in sight. They were labelled Mount Zeehan and Mount Heemskirk. Additional features of the terrain might have been named, except that the unsettled weather and rough, choppy seas made a more thorough exploration along the island's east coast too

---

21. Words in this sentence taken from A. Curnow, 'Landfall in Unknown Seas', in A. Curnow, *Collected Poems, 1933-1973* (Wellington: A.H. & A.W. Reed, 1974), pp. 136-39.
22. B.N. Hooker, 'Towards the Identification of the Terrestrial Globe Carried on the *Heemskerck* by Abel Tasman in 1642-43', *The Globe* (Journal of the Australian and New Zealand Map Society) 79 (2016), pp. 31-32; H. Gerritsz, *Mar del Sur. Mar Pacifico*, Bibliothèque nationale de France, Département des Cartes et plans, cote GE SH ARCH-30.
23. Hooker, 'Towards the Identification of the Terrestrial Globe', p. 31.
24. Ibid., pp. 33-34; Sharp, *The Voyages of Abel Janszoon Tasman*, p. 81.
25. Bennett, 'Van Dieman, Tasman and the Dutch Reconnaissance', p. 68.

risky. Tasman found shelter in a bay (which he named Storm Bay) and waited for the violent weather to abate. Finally, on 1 December, with the storm having passed, the ships dropped anchor at Green Island (just over three kilometres off the mainland). Tasman claimed the island for the Netherlands; but there was insufficient water there to supply the vessels, and so the decision was made to leave the area altogether, and sail further east still – again into a vast oceanic region where no European had previously explored.

There is a sense of impatience at this point in Tasman's proceedings. Perhaps the fatigue and frustration of having travelled nearly 14,000 kilometres since leaving Batavia on this exploratory mission (with no obvious commercial benefits having emerged thus far) was beginning to exasperate him, particularly as almost the entirety of his career to date had been fashioned around trade. Perhaps, too, he was disillusioned with his only 'discovery' to date having been an inhospitable, apparently desolate island that seemed to promise so little that he did not even venture to circumnavigate it. It is also possible that Tasman held some vague hope, however diminished, that a territorial prize still awaited him somewhere over the horizon to the east. Up until this point, though, the only item of (questionable) value that had been extracted from this great voyage of discovery was an outline map of the west coast of Tasmania. However, as he left the island for whatever might lie ahead, his optimism seems to have gradually wilted, and from this point, the expedition seemed to take on an air that was much more perfunctory than passionate.[26]

'Staete Landt'

'Again', Tasman jotted dispiritedly in his journal, as he ordered the ships to leave Tasmania on 5 December 1642 and once more sail eastward across unknown distances of ocean.[27] The vessels traversed what would become known as the Tasman Sea[28] until midday on 13 December, when 'a large and high land' was sighted on the horizon. Piecing together all that was known about the adjacent parts of the world, Tasman deduced (wrongly as it turned out) that this was most likely the western extremity of Staten Landt – a potentially enormous landmass which was connected to South

26. See Heeres and Coote, *Abel Janszoon Tasman's Journal*, p. 112.
27. Ibid.
28. H. Rotschi and L. Lemasson, 'Oceanography of the Coral and Tasman Seas', *Oceanogr. Mar. Biol. Ann. Rev.* 5 (1967), 49-98.

A view of the Murderers' Bay, as you are at anchor here in 15 fathom
(Isaac Gilsemans, 1642)
Nationaal Archief, The Hague

America[29] – and accordingly appended this name to the territory. The vessels were sailing towards the west coast of what would later be known as the South Island of New Zealand, in an area between Hokitika and Ōkārito. The following day, Tasman wrote of approaching the country:

> We were about two miles from the land. It was a very high, double land, but from the thick clouds we could not see the tops of the mountains. We shaped our course northerly, and so close that we could see the surf breaking on shore. In the afternoon, about two miles from shore, we sounded in 55 fathoms, sticky, sandy ground. It was calm. Towards evening we saw a low point, about three miles from us northeast by north. We drifted quietly towards it. In the middle of the afternoon we sounded in 25 fathoms, sticky, sandy ground.

29. Staten Landt was first encountered by Jacob Le Maire and Willem Schouten in 1615. W. Schouten, *Journal ou description du merveilleux voyage de Guillaume Schouten, Hollandois natif de Hoorn, fait es années 1615, 1616 & 1617* (Amsterdam: Harman Janson, 1619), p. 24.

> We sailed along quietly the whole night, the current setting in from the west-north-west. We neared the land till within 28 fathoms, good anchor-ground; it still being calm, and not to go nearer the land we anchored ... and waited for the land-wind.[30]

The coastline that confronted these Dutch sailors was forbidding. It was lined, as far as the eye could see, by a strip of black sand that was relentlessly being pounded by ocean swells crashing into it. From there, the land rose steeply and unevenly, and was densely draped in a primordial podocarp forest. Further in the distance, this dull-green canopy thinned, eventually giving way to a snow-capped alpine range, cut into innumerable ridges and crevices.

From this location, the ships headed north, keeping close to the coast and slowly sketching what would become the first map of part of New Zealand. Various details were mixed into this cartographic cauldron, including seabed depths, the topography of the terrain that could be seen from the water (and how it altered depending on how far a vessel was from shore), places of refuge, tidal movements, natural hazards and the curvature of the coast. There were navigation manuals with which all mapmakers affiliated to the company – including Tasman and his fellow officers – were obliged to comply and which imposed uniform standards and a strict emphasis on precision.[31]

However, mapmaking was still very much a fledgling science in this era. Longitudinal measurements, in particular, remained little more than informed estimates, resulting in false proportions often appearing on maps of the period, such as the elongated arc of northern Queensland that featured on Tasman's map of Australia.[32]

The equipment that the Dutch and others depended on when putting their discoveries to paper in the first half of the 1600s inevitably made their works at best impressionistic in parts. Sextants (along with marine chronometers), which were used to calculate distances with reasonable

---

30. A. Tasman, in T.M. Hocken, 'Abel Tasman and His Journal', *Transactions and Proceedings of the Royal Society of New Zealand* 28 (1895), p. 128.
31. Zandvliet, *Mapping the Dutch World Overseas*, pp. 1435-36.
32. A. Tasman, 'Australia and New Zealand [cartographic material]: from the original map made under the direction of Abel Tasman in 1644 and now in the Mitchell Library, Sydney / this facsimile was drawn by James Emery in 1946' (Sydney: Trustees of the Public Library of New South Wales, 1946), ref. rbr293624.

precision by measuring the relative position of astronomical objects to the horizon, did not come into use until around a century after Tasman's voyage to New Zealand. Instead, mariners relied on the astrolabe and the cross-staff, which were useful principally for calculating latitude. The challenge of measuring longitude with any exactitude was thought to be impossible (even at the end of the eighteenth century) because of the unlikelihood of timepieces being invented that would be sufficiently accurate. Tasman depended on a sand-glass to track time with sand passing through a small aperture between two glass bulbs – a process that was measured to take a specified period to complete. (The term 'glasses' was often used instead of hours as a measurement on ships in this period.)[33] The other piece of equipment that was indispensable to explorers in this era was the telescope but, as with other navigational technology, it was still in its infancy. Through these slightly opaque lenses, Tasman recorded both in Tasmania and New Zealand that members of the indigenous populations in those territories were above average in size.[34] It is possible, however, that such assessments were the result of poorly refined glass in the telescopes that distorted their magnification[35] – a fact that may have also affected the accuracy of some of the topographical drawings produced on this expedition. In a way, it also serves as a metaphor for the disfigured European gaze at the non-European world: a sketchy visual narrative produced by squinting for a glimpse of the unknown, which is first made known through cartography, even if the dimensions are awry at times.

* * *

Four days after sighting New Zealand's coast, Tasman led his ships as far as Cape Farewell – the northernmost point of the South Island – where the vessels anchored. A thin finger of smoke rising from a spot beneath the forest canopy close to the coast suggested that the location was inhabited by humans, but as the sun rose on 17 December, no occupants of the land had yet been seen by the crews. That evening, the glow of lights from fires on the shore were detected by the Dutch sailors and, as they kept watch, they observed a group of Māori boarding two *waka* (canoes) and paddling them towards them. Some of those in the *waka* called out in the direction of the ships, but the language barrier

33. Hocken, 'Abel Tasman and His Journal', pp. 117-22.
34. Sharp, *The Voyages of Abel Janszoon Tasman*, p. 110.
35. G. Anderson, *The Merchant of Zeehaen: Isaac Gilsemans and the Voyages of Abel Tasman* (Wellington: Te Papa Press, 2001), p. 103.

left their guttural messages unknown to Tasman's crew. At least one of the Māori then began blowing a 'strange' instrument (most likely a *pūtātara*, made from a conch or triton shell).[36] One of the Dutch sailors responded to this overture with a trumpet fanfare, which concluded the first known, cross-cultural exchange between Māori and Europeans. It was also the commencement of the long catalogue of misunderstandings and misplaced efforts at bridging a vast cultural gulf that separated the two peoples.

Unlike their previous landfall, in Tasmania, this time, the officers of the ships sensed that more caution was needed before going on shore. It was an apprehension that was soon justified. The following morning, several *waka* (up to 22 of them)[37] set out from the beach in the direction of these visitors, reaching their destination just as a group of Dutch sailors was rowing from one ship to the other. The Māori paddlers raced 'furiously' towards this boat and, when close enough, attacked some of the seamen, killing three almost immediately and fatally injuring a fourth.[38] Tasman ordered a barrage of fire against the fleeing assailants and then promptly made the decision to leave the area, abandoning any hope that he could get water or other supplies there.[39] The most lasting cartographical impression this event left was the naming of the location by Tasman: De Moordenaars Baay – Murderers' Bay[40] (incidentally, just one of six names the place has been known as since this time).

One of the limitations with maps was that they did not alert their users to the sort of dangers Tasman and his crews had just faced. Illustrations were therefore used on occasion to augment the record of these exploratory journeys but, without the need for the sort of precision cartographers invested in their maps, their purpose could afford to be more polemical. Crucially, though, these pictures were created specifically to be seen in conjunction with maps of the areas in question, and so added a moral tinge to the charts that were produced on the voyages. The response of Gilsemans, the expedition's draughtsman, to the four murdered Dutchmen was to depict the confrontation with

---

36. B.J. Slot, *Abel Tasman and the Discovery of New Zealand* (Amsterdam: Otto Cramwinckel, 1992), p. 61.
37. Sharp, *The Voyages of Abel Janszoon Tasman*, p. 123.
38. Slot, *Abel Tasman and the Discovery of New Zealand*, pp. 62-63.
39. Anderson, *The Merchant of Zeehaen*, p. 92.
40. A. Tasman, in J. Burney (ed.), *A Chronological History of the Discoveries in the South Sea or Pacific Ocean* (London: Luke Hansard, 1803), pp. 63, 75.

Māori in a manner that removed the boundaries between representation and vilification. The resulting engraving was cartographically precise, even though it was simultaneously a work of ethnographic and anatomical errancy. In the background of *A* view *of the* Murderers' *Bay, as* you *are at* anchor *here in 15 fathom*,[41] Gilsemans painstakingly illustrated the terrain along this portion of the coast, almost as if hoping that such accuracy would imply his portrayal of Māori was similarly representative. In the middle ground of the picture are the two Dutch ships, next to which is a key that indicates some of the elements assembled in the picture. It is the image in the foreground, however, to which the viewer's eye is first drawn: an out-of-scale canoe, occupied by about a dozen Māori and trailed by a fleet of other *waka* from the coast. Instead of literal representation, however, Gilsemans delivered evil caricature. The paddlers appear almost otherworldly, with grossly enlarged heads and unusually positioned bodies. This is as much a conjuring of The Other, excavated from the artist's imagination, rather than any attempt at an honest likeness. The only bright spot (literally) in the picture is the small Dutch flag in the centre of the work, which is so audacious because it is coloured and thus seems to protrude from the surrounding sea of black and white. The decision to add colour solely to this flag may have been made by the publisher rather than Gilsemans but, regardless, it is a vibrant hint at which nation wishes to signal that it exercises dominion in this scene.

For all its rhetorical force, though, Gilsemans' image is still part of the earliest corpus of cartography relating to New Zealand. Once the viewer had finished gazing at those Māori who dominate the scene, he or she could scrutinise the topography of this part of the country, which the image much more faithfully represents. Gilsemans may not have been exclusively responsible for the final version of this picture, however. Even though the illustrations in Tasman's journal were attributed to him, it is probable that Frans Visscher had a hand in them.[42] It was Visscher's thick nib that also outlined in ink part of the coast of the South Island, stretching to the lower North Island, producing what became the first map plotting a portion of New Zealand.

---

41. I. Gilsemans, *A* view *of the* Murderers' *Bay, as* you *are at* anchor *here in 15 fathom* (1642), National Library of New Zealand, Alexander Turnbull Library, ref. PUBL-0086-021.
42. W. Eisler, *The Furthest Shore: Images of Terra Australis from the Middle Ages to Captain Cook* (Cambridge: Cambridge University Press, 1995), pp. 83-85.

From Murderers' Bay, the fleet, with its crew despondent after the recent killing of four of their colleagues, sailed east to what would become known as D'Urville Island. At this point, the weather began to deteriorate. Molten grey clouds poured out over the sky, bringing heavy rain, squally winds and heaving swells. It was not until 26 December that the storm swept away to the Pacific Ocean and the vessels were able to continue with their exploration. Crossing the body of water separating the country's two main islands, Visscher suspected that there was a passage through, but pressure of time meant that the fleet had to continue sailing north-west, rather than probing further east to plot the strait.

By the following day, the ships were around 230 kilometres west of Awakino. From there, Tasman altered course, this time sailing back towards the coast 'in order to ascertain whether the land we had previously seen in 40° extends still further northward, or whether it falls away to eastward', and on 29 December, as they once again drew closer to the coast, they caught a sight of the extinct volcanic cone of Mount Karioi (just south of present-day Raglan), which at 750 metres was one of the more prominent landscape features visible from offshore on this leg of the voyage. From there, the ships continued hugging the west coast of the North Island, passing close to Cape Reinga on 4 January 1643. Two days later, Tasman steered his vessel towards Manawatāwhi (Three Kings Islands – 60 kilometres north-west of Cape Reinga), with the aim of replenishing his supplies of water and food, but the sight of Māori on the island deterred him and he again decided to avoid the risk of going on shore. Instead, he continued to sail north, away from New Zealand, and onto the final leg of his expedition.

Batavia, Amsterdam and Commercial Decisions

Where we might have expected elation, there was just exhaustion. After ten momentous months at sea, punctuated by short periods of coastal mapping and a fatal encounter with 'barbarians',[43] Tasman's two weather-worn ships staggered back into the port at Batavia on 15 June 1643. As the vessels docked, he was met not with fanfare or popular jubilation, but by wharf workers who were indifferent to the feats of this expedition, along with a few local officials who turned up to carry out their perfunctory duties. Such was the Dutch East India Company's

43. P. Moon, 'Shades of the Savage in Colonial New Zealand', *Journal of Colonialism and Colonial History* 18, no. 2 (2017), pp. 1-10.

commercial preoccupation in this period that Tasman's discoveries would be evaluated not for any contribution that they made to Europe's geographical or anthropological knowledge of the world, but for more mercantile motives.

By the nineteenth century, the discovery of new lands by European powers was often a cause for chest-thumping, flag-waving, patriotic triumphalism. Imperial prowess was increasingly measured by the area of territories being claimed, followed by assessments of their strategic importance and whatever opportunities they presented to extend European virtues to the 'savages' living in those locations. In the first half of the seventeenth century, though, the Dutch East India Company took a more calculating view of its colonial incursions. Territorial claims were upheld only if there was sufficient financial return to be had. The appetite for exploration in particular regions quickly diminished once it became clear that there was no commercial advantage in the lands that the company was mapping. Trading opportunities, scarce resources, or navigational short-cuts were what company officials were really interested in. Anything else was just there to satisfy latent curiosity.

Several of those on board the *Heemskerck* and *Zeehaen* during this epic expedition had produced cartographic and other material detailing routes travelled and territories encountered. The first task for the company, working with the ships' officers, was to assess the intelligence that had been accumulated during the voyage, and establish how they might profit – in every sense – from it. However, despite the lengthy ordeal of the journey, there was not a huge quantity of written material for officials to sift through. Tasman's records were cartographically routine and his journal accompanying them was sparse when it came to details of the places he had visited.

Four days after Tasman's return to Batavia, the local company directors wrote to their head office in Amsterdam, outlining the purpose and accomplishments of the expedition. In order to justify to their superiors the money they had spent on this voyage, these officials tried to squeeze what few drops of good news they could from the reports produced during the expedition. Tasman and his crew had been promised a financial reward if they succeeded in discovering resources or commercial opportunities, or if they mapped new passages for navigation. However, although the Batavian officials noted that 'no treasures or matters of great profit have as yet been found', and that further exploration would be required to determine whether there was any additional value in the territories and routes that Tasman charted, the voyage had been so substantial that the crews and officers of the ships would be rewarded regardless.[44] This assessment, along with a summary of the expedition, was then despatched on the next ship bound for the Netherlands.[45]

As with the mechanics of navigation, Dutch colonial bureaucracy was still in its infancy in this period. There was no standardised system of filing or duplicating the various forms of documentation produced on ships, and neither was there a consistent reporting regime in place between Batavia and Amsterdam. Instead, to avoid a mounting clutter of paperwork, what was judged to be valuable was sometimes stored and sometimes copied and sent to the relevant company officials in other parts of the Dutch Empire, while those documents deemed to be of less consequence eventually tended to disappear into an administrative oblivion (and in a handful of cases, resurfaced decades, or even centuries, later in various private collections). In some instances, however, there were leaks in the conduits through which intelligence flowed, leading to information being shared more widely than it otherwise would have been. The other feature of this fledgling bureaucracy was the absence of a full system of cataloguing the documents that passed through it. As a result, what happened to the intelligence amassed in Batavia (and elsewhere within the company's dominions) cannot now be fully reconstructed. Not only are many of the records lost to history, but even the existence of some of these documents can only be speculated on. What seems most probable is that, in early 1643, Tasman personally supervised the preparation of all the documentation produced during his voyage and most of these records eventually ended up in Amsterdam.

With not even a hint of gold or silver being mentioned anywhere in Tasman's journal of his lengthy expedition to catch the eye of his employers, company officials soon abandoned any detailed interest in the rest of what he had recorded about the voyage. Maybe, though, there was more to find. Perhaps there was some potential strategic benefit or navigational advantage that would surface after a more thorough exploration of the various locations that Tasman identified. After mulling over the documents produced during the expedition, company officials in Batavia concluded that a further voyage would be advisable, on the basis of a 'well-founded hope that something profitable will ultimately

---

44. Heeres and Coote, *Abel Janszoon Tasman's Journal*, p. 144.
45. I. Burnet, *The Tasman Map: The Biography of a Map* (Sydney: Rosenberg Publishing, 2019), p. 190.

turn up'.[46] Accordingly, they wrote to the directors in Amsterdam on 22 December 1643, offering an appraisal of Tasman's recent journey in a letter that was partly an explanation and partly hidden plea for support for a second expedition:

> Several new lands and islands have ... been discovered in the south, besides which they affirm that they have found an open passage into the South Sea to get to Chili. ... [W]e also send ... the daily registers kept by the aforesaid Tasman and the Pilot-major Francois Jacobsen Visscher, the said registers pertinently showing the winds and the courses held, and faithfully delineating the aspect and trend of the coasts, and the outward figure of the natives, etc. We have, however, observed that the said commander has been somewhat remiss in investigating the situation, conformation and nature of the lands discovered, and of the natives inhabiting the same, and as regards the main point has left everything to be more closely inquired into by more industrious successors. It also appears that in running to southward from the island of Mauritius, they did not sight any land until they had come to the 49th degree; but thence going eastward they finally got into the South Sea to the south of the South-land. Now, that in this latitude there really is a passage to Chili and Peru, as the discoverers stoutly affirm, we are not prepared to take for granted, since, if they had run a few more degrees to the south, they might not unlikely have come upon land again, perhaps even upon the Statenland (thus named by them) which they had left south of them, and which may possibly extend as far as Le Maire Strait, or maybe even many more miles to eastward. All this is mere guess-work, and nothing positive can be laid down respecting unknown matters.[47]

It did not matter that Tasman's reputation was slightly muddied in this report. Its purpose was to justify to the directors in Amsterdam the need for another expedition (and presumably, to gain authorisation for

46. A. van Diemen to Noble, worshipful, wise, provident and very discreet gentlemen, The Dutch East India Company, Amsterdam, 22 December 1643, in ibid., p. 145.
47. Ibid., pp. 144-45.

the expenditure such an undertaking would require). This letter is even more significant, though, because it provides a crucial link in the chain of evidence confirming that the cartographical information Tasman and his colleagues produced about New Zealand reached the Netherlands in 1644. There are no signs that this was done in any carefully curated manner but, rather, that most of the available documentation for the expedition was probably bundled together and shipped to Amsterdam. With no obvious discoveries of value contained in them, predictably, the company's interest in these maps and journals soon dried up, but not before they were handed over to the company cartographers, whose first job was to extract any information that could be used to enhance their charts of this part of the world.

Chapter 4

'The Intelligence Empire': Seventeenth-Century Dutch Exploration of the South Pacific

The Nature of VOC Information Gathering

During that long twilight of their empire, British writers relished in the glow of their nation's imperial triumphs. Britain had held 'dominion over palm and pine', as Kipling disarmingly put it and, by the end of the nineteenth century, it had evolved into a force of global benevolence. The Empire stood for 'protection for weaker races, justice for the oppressed, liberty for the down-trodden'[1] and was guided by the noble aim 'to implant – slowly, prudently, judiciously – those ideas of justice, law, humanity, which are the foundation of our own civilization'.[2] This is how empires can finish, in a patriotic delirium that clouds both their view of the more unpleasant dimensions of colonisation as well as of their own accomplishments.

Not all imperial powers saw themselves in the same light, though. By the beginning of the twentieth century, some lesser European states were frantically engaged in an unedifying 'scramble for colonies' – attempting to secure a small toehold usually in an otherwise neglected corner of Africa that remained unclaimed by any other European nation. These

1. Alfred Harmsworth, quoted in W.L. Langer, *The Diplomacy of Imperialism, 1890-1902* (New York: Alfred A. Knopf, 1965), p. 84.
2. John Morley, quoted in D. Gilmour, *The Ruling Caste: Imperial Lives in the Victorian Raj* (London: Pimlico, 2005), pp. 22-23.

late starters in the imperial race were endeavouring to mimic the sort of imperial success that had helped to elevate the fortunes and status of countries such as Britain. However, they tended to see territorial possessions as an end in themselves, to be acquired regardless of the moral costs involved, or even of any certain financial gain.

Both these approaches to imperialism – either as benign paternalism or avaricious interference – have sometimes tended to overshadow perceptions of earlier periods of Europe's colonial expansion. Certainly, competition always existed between imperial powers, but in the case of the British and the Dutch in the seventeenth century, there were only relatively short periods of serious rivalry which punctuated otherwise long stretches of cautious coexistence.[3] The notion of empire in this period differed dramatically from its later incarnations. Very roughly, what preceded a military and political empire – of the sort familiar in the nineteenth century – was a territorial one; and what preceded a territorial empire was a commercial one. However, before all these forms – at the genesis of imperialism – existed what could be described as an 'intelligence empire': a category into which the Dutch exploration of the South Pacific in the seventeenth century fits.

There are several traits which typified the 'intelligence empire' phase of European imperialism. First, the intelligence gathered tended to remain confidential only for short periods, after which it began to appear in books and documents that were circulated relatively freely. Second, the type of information collected tended to be 'physical' scientific data relating to geography, oceanography, botany and cartography, rather than anthropological details about the peoples and cultures in the territories that were being explored (that would come later). This reflected the priority at this time of getting to a location over understanding the inhabitants of that location. To a degree, there was also a sense of cultural superiority woven into this approach, in which Europeans excluded the possibility of encountering any civilisation that would be regarded as being even vaguely equivalent, let alone better than their own. Thus, the accounts of the indigenous inhabitants and cultures of the territories being explored could be left to others later on to document.[4]

---

3. K.N. Chaudhuri, *The English East India Company: The Study of an Early Joint-Stock Company 1600-1640* (London: Frank Cass & Co., 1965), p. 49.
4. P. Moon, 'From Tasman to Cook: The Proto-intelligence Phase of New Zealand's Colonisation', *Journal of Intelligence History* 18, no. 2 (2019), p. 254; Bayly, *Empire and Information*, p. 3; M.H. Edney, *Mapping an Empire: The Geographical Construction of British India, 1765-1843* (Chicago: University

Another defining characteristic of this 'intelligence empire' phase of colonisation is the relative absence of a systematic way of dealing with the knowledge generated from voyages of discovery, along with a surprising degree of cooperation and sharing between colonial powers.[5] It can be useful to view intelligence in this era is a form of currency with a fluctuating value. As the case of Tasman's mapping of New Zealand illustrates, the cartographical information he gathered on that voyage was initially kept confidential, as a matter of course, but then ended up being shared, and later formed the basis of other maps, which were subsequently treated as being commercially or politically valuable when they began to be used for new purposes, until they too became 'shareable'.[6]

The way in which information gathered by Tasman in 1642 later migrated its way through Europe reveals that this sort of intelligence enjoyed a form of autonomy. It was almost as though once it had come into being, it took on a life of its own and functioned similar to a living organism: it was responsive to its environment; it could modify itself according to its surroundings; and to an extent was capable of reproduction and evolution.[7] The mapmaking studios in mid-seventeenth-century Amsterdam, the cartographical publishers scattered across Europe busily producing their own maps and accompanying narratives, the coffees and chats between officials and diplomats representing different nations at which information slipped from one person to another with the ease of gossip, and the ingratiating enquiries from merchants, agents and shipowners about the latest discoveries that might affect their businesses, were all ultimately beyond the ability of any one organisation – even one as seemingly omnipotent as the Dutch East India Company – to control completely.

Various metaphors have been deployed to describe this imperial organism. Some historians have depicted it as a web, with a complex interlacing of connections between all the locations, groups and individuals involved,[8] while others have portrayed it as a 'world system' of

of Chicago Press, 2009), p. 319; G.V. Scammell, *The First Imperial Age: European Overseas Expansion c. 1400-1715* (London: Routledge, 2003), p. 27; M. Castells, *The Information Age: Economy, Society and Culture: Volume 1: The Rise of the Network Society* (Oxford: Blackwell, 1996), chap. 1.

5. Moon, 'From Tasman to Cook', p. 255.
6. Ibid.
7. Ibid.; C. Bayly, 'Knowing the Country: Empire and Information in India', *Modern Asian Studies* 27, no. 1 (1993), p. 5.
8. T. Ballantyne, *Webs of Empire: Locating New Zealand's Colonial Past* (Wellington: Bridget Williams Books, 2012), pp. 25-26.

encroachment and exploitation, which extended 'in chain-like fashion' from the centre to the periphery.[9] However, whatever imagery or simile has been applied in an effort to conjure up the essence of imperialism, they all share two points of agreement: that the systems were enormously complex; and that they began with the accumulation of information.

* * *

In an effort by the Dutch East India Company to bring some order to the growing piles of intelligence amassing in its headquarters, from the early 1620s, the chief of its Hydrographic Office was instructed to take sole responsibility for all the journals, maps and drawings arriving in Amsterdam from all the locations in the company's empire. His role was to condense this information and to make any corrections to maps arising from new intelligence that had been gathered.[10] However, these changes could only be officially made once the formal authorisation from the company's directors had been given. Mapmakers employed by the company therefore had to report to the directors every six months with a summary of the alterations that they had made to charts. 'Secrecy was and remained the watchword', as fears of foreign competition became more acute in this period,[11] and strict prohibitions were initially imposed on the printing or publishing of any material without the explicit approval of the company's directors. Indeed, those producing charts had to take an oath promising that they would maintain full confidentiality about their work, which extended to not talking about their activities with anyone outside the company.[12]

* * *

In the autumn of 1632, at the age of around 51, Hessell Gerritsz – Amsterdam's master mapmaker – died. However, the company did not find itself suddenly in the dark because of his death. In their wisdom,

9. A.G. Frank, *Latin America: Underdevelopment or Revolution: Essays on the Development of Underdevelopment and the Immediate Enemy* (New York: Monthly Review Press, 1970), pp. 7-8; I. Wallerstein, 'The Rise and Future Demise of the World Capitalist System: Concepts for Comparative Analysis', *Comparative Studies in Society and History* 16, no. 4 (September 1974), 387-415.
10. Keuning, 'Hessel Gerritsz', pp. 56-60.
11. Ibid.
12. Resolutions by the Chamber of Amsterdam, no. 360 (9 December 1632), Algemeen Rijksarchief, Koloniaale Afdeling, in Schilder, 'Organization and Evolution', p. 62.

the directors had long ago erected the architecture of the company in a way that no single individual's presence (or absence) could derail its ambitions in any significant way. Prospective replacements were always lined up to fill gaps caused by deaths or departures. In the great meeting halls of its headquarters, and also closeted away in its offices, chambers and cloisters – all connected by a labyrinth of dark, wood-panelled passageways – the company's apparatus continued, uninterrupted by the comings and goings of individuals – even those who had been as vital as Gerritsz. Inside this grand building, to the smell of smouldering tobacco wafting from Gouda pipes, and logs burning in its innumerable fireplaces, senior company officials moved quickly to ensure that its vital mapmaking enterprise continued.

In 1633, Willem Janszoon Blaeu, now 62 years old, was put in charge of the four assistants that had been working for Gerritsz.[13] Five years later, when Willem Blaeu died, his son Joan Blaeu (who had been trained as a cartographer by his father) was appointed by the company to succeed him 'on conditions and at a remuneration which shall be equal to his late fathers'. It would be a role he would occupy for the next 35 years.[14]

It was early on in Joan Blaeu's career that he received Tasman's maps from the 1642-43 Pacific expedition – at a time when a noticeable shift in the culture of the company was taking place (albeit at glacial slowness). Whereas in the early 1630s, the directors clamped down tightly on every scrap of intelligence that was flowing into its headquarters, as though their fortunes depended on complete secrecy, by the 1640s, this attitude was changing. Although the younger Blaeu had committed himself to the same company-mandated strictures of secrecy that bound his predecessors, such prohibitions were beginning to be honoured more in the breach than the observance. His personal frustration lay in the fact that he was unable to reconcile the increasing quantity and quality of the cartographic material he was receiving with the requirement to keep it concealed and, in some cases, consign it to the enforced forgetfulness of the company archives.[15] As a seasoned mapmaker, he knew that

13. Ibid.
14. Ibid., p. 65.
15. As an indication of the vast scale of records generated by the company, it has been estimated that surviving papers relating to it comprise an estimated 25 million pages of historical records, stored over four kilometres of shelf space. J. Lucassen, 'A Multinational and Its Labor Force: The Dutch East India Company, 1595-1795', *International Labor and Working-Class History* 66 (2004), p. 13.

much of the material that came across his desk probably had negligible commercial value to the company, but was nonetheless important when it came to adding to the body of knowledge about the geographical form of the world. The solution, he convinced himself, was to allow a small quantity of the more useful intelligence to form the basis of a commercial publishing enterprise which he formulated.

His method was cautious but shrewd. First, he increased the amount he charged the company for the sets of charts he made for their fleets. (As these charts were all hand-drafted rather than printed, he could control both their manufacture and distribution.) Then, with the additional revenue he obtained from this price-hike, he was able to produce a substantial volume of additional maps, manuals and globes, which in turn generated even greater profit.[16] Technically speaking, he was still abiding by the requirement to keep all the material he issued within the confines of company employees. However, the sheer quantity of his output was such that its spread beyond the company's reach was more a case of when rather than if. This suggests that the company's earlier apprehension about others obtaining cartographical information on its colonial possessions was receding slightly by this time, which explains its tacit preparedness to sanction Blaeu's more entrepreneurial approach to his mapmaking work. As a result, information about New Zealand was on the cusp of trickling out from the bounds of the company's headquarters and its network of mariners, and into much wider and unbridled circulation.

At first glance, this might give the impression of an organisation whose authoritative hold on its operations was gradually loosening – as though it was not now of that strength which had formerly led to its expansion. However, such a view would be a misunderstanding of the essential nature of the Dutch East India Company. In this period, at least, its successes derived less from foresight or structural rigidity and more from 'piecemeal engineering to remedy design flaw[s]'.[17] So, provided that Blaeu's profiteering, and his calculated disregard of some of the more obsessive requirements for secrecy, did not impinge on the company's profitability, then there was no design flaw that required remedy. Moreover, as the company's trade routes were by this time already well established and well-known anyway, the ongoing insistence on secrecy in most cases was becoming increasingly redundant.

---

16. Zandvliet, 'Mapping the Dutch World Overseas', p. 1439.
17. Gelderblom, de Jong and Jonker, 'The Formative Years of the Modern Corporation', p. 105.

By the latter part of the 1640s, the company's directors, both in Batavia and Amsterdam, seemed to have concluded that, in the foreseeable future, New Zealand would be of no commercial, strategic or political value to them. More generally, in 1645, company officials in Amsterdam stated that they could not 'anticipate any great results from the continuation of such discoveries, which besides entail further expenditure for the Company'. Instead, their preference was for maintaining the existing territories and resources that they exercised dominion over.[18] From this point onwards, as intelligence about New Zealand went from the private to the public sphere in Europe, there would be less control than ever before about what would happen to that information, along with a much more sizable audience now having access to it.

This slightly more open approach to some of the cartographical material that the company had commissioned and collected was mirrored in Batavia. Officials in this company-controlled state were beginning to place a more pragmatic value on some of the intelligence reaching them from mariners such as Tasman. Without a clear commercial or strategic justification to maintain the secrecy of the maps and coordinates that its sailors produced, the company directors in Batavia began to see the futility in concealing these documents in perpetuity. At no previous time had circumstances been so ripe for the Dutch discoveries in the South Pacific to make their way into popular knowledge in Europe.

The English East India Company

As Portuguese and Spanish power in South East Asia crumbled away in the early seventeenth century, the Dutch were not the only Europeans keen to fill the commercial void. In 1600, a Royal Charter was granted to 'the Governor and Company of Merchants in London trading to the East Indies',[19] which led to the formation of the English East India Company (EIC). What started off as a tentative and speculative venture to 'make trade ... by buying or barteringe of suche goodes wares jewelles or merchaundize as those Ilandes or Cuntries may yeld or afforthe'[20]

18. O.H.K. Spate, *Monopolists and Freebooters [The Pacific Since Magellan: Volume 2]* (Canberra: Australian National University Press, 1983), p. 51, in Bennett, 'Van Diemen, Tasman and the Dutch Reconnaissance', p. 74.
19. J.R. McCulloch, *A Treatise on the Principles, Practice, & History of Commerce* (London: Baldwin & Cradock, 1833), p. 120.
20. The Court Records of the East India Company, 24 September 1599, in H. Stevens, *The Dawn of British Trade to the East Indies as Recorded in*

grew erratically but determinedly in the following decades.[21] The EIC set up what was termed a 'factory' (initially a trading centre for pepper) in Bantam, which neighboured Batavia. Bantam was, in every respect, the poor cousin of its Dutch counterpart. It was smaller in scale, starved of sufficient capital, mediocrely managed and had none of the great civic planning that had allowed Batavia to flourish.

Normally, two commercial, colonising powers, squeezed together in competition on a small island, would be a recipe for friction and, indeed, there were several tense periods between the English and Dutch in Java. However, for the most part, relations between the two nations were governed more out of pragmatism than misguided patriotism (despite the fact that there was an air of bullishness from the English upstarts in the region, with one official warning that 'although the Hollanders threaten any who do but peep into these parts, they will be better advised than to proceed with open force to make the English their enemies').[22] Ultimately, what bound the Dutch and the English in the early decades of the 1600s – apart from a few episodes of fighting[23] – was the lingering presence in the region of their mutual Catholic enemies. Nonetheless, as this threat receded, so too did the prospects for long-term cohabitation between these two Protestant powers and, in 1682, the English were finally squeezed out of their trading post as a result of a political deal forged between the Dutch and the local Sultan.

* * *

Just as armies march on their stomachs, so do empires advance on their bureaucracies. From its inception, the English East India Company's Bantam outpost generated and stockpiled paperwork which detailed its commercial transactions and, occasionally, other aspects of its operations. These records were compiled for the immediate needs of the business rather than for the curiosity of later generations. It was not until the 1890s, when Sir George Birdwood rolled up his sleeves and began to

---

the Court Minutes of the East India Company 1599-1603 (London: Henry Stevens & Son, 1886), p. 5.

21. K.N. Chaudhuri, *The English East India Company: The Study of an Early Joint-Stock Company 1600-1640* (London: Frank Cass & Co., 1965), p. 4.
22. English East India Company, Court of Committees, 21 April 1615, in ibid., p. 48.
23. W. Milburn, *Oriental Commerce: Containing a Geographical Description of the Principal Places in the East Indies, China, and Japan* (London: Black, Parry & Co., 1813), p. xv.

burrow his way through this forbidding jungle of documents, that some insights emerged into the EIC's workings in the seventeenth century. Even then, though, most of the material that Birdwood assembled was fairly pedestrian in nature, with inventories of prices and quantities of goods dominating the documentation. However, buried deep in the midst of these aged archives, one fragment of a sentence from a letter dated 10 January 1644 represents an axiomatic point in New Zealand's history. The fragment was contained in a letter sent from an official in Bantam to his superiors in London and reads: 'A draught of the South Land lately discovered'. It was appended to 'a rough sketch, much damaged, and only kept together by being backed with goldbeater's skin'.[24] That sketch was of Tasman's map of his exploration of a portion of the South Pacific.

Almost certainly, this was the first occasion where information about New Zealand's existence and its location went from Dutch into English hands. Whether this was done with the approval of the Dutch authorities in Batavia is unknown, but highly unlikely. In early 1644, when this letter and map were sent to London, the Dutch VOC cartographers in Amsterdam would themselves only just have received the bundle of records relating to Tasman's recently completed expedition and would have been processing them under the secrecy that blanketed all their work. It is therefore exceedingly improbable that Batavian officials would have been so free with this intelligence, knowing that no final determination about its value had yet been made by their superiors in the Netherlands. Most probably, this intelligence drifted into the realm of the English East India Company via unnamed Dutch merchants or sailors who passed on snippets of information about Tasman's discoveries when visiting the English trading post in Bantam (quite possibly, some of them had been among the crew on that expedition). In that January 1644 letter, the English official based in Bantam provided the following details which he managed to assemble from his informants:

> The Dutch have lately made a new discovery of the South Land in the latt[itude] of 44 d[egrees] and their longitude 169, the draught whereof is herewith sent. They relate of a gyantlike kinde of people there, very treacherouse, that tore

---

24. Contained in the 'D.P. [Damaged Papers] Volumes', no. 2,397, Bundle A, marked 'Miscellaneous', covering the period 1623-1708, in G.C.M. Birdwood, *Report on the Old Records of the India Office, with Supplementary Note and Appendices* (London: W.H. Allen & Co., 1891), p. 77.

> in peeces lymbemeale [i.e. limb by limb] their merchant, and would have done them further mischeife, had they not betaken them to their shipps. They make mention alsoe of another sort of people about our stature, very white, ruddy, and comely, a people gentle and familiar, with whome, by their owne rellations, they have had some private conference. We are tould that the Dutch Generall [i.e. Governor-General] intends to send thither againe and fortifie, having mett with some thing worth the looking after. They have discovered a second Mauritius, some 80 leagues to the northwards of the other, whose draught [missing] is likewise herewith sent you; both the one and the other not to be had or procured from them but by extraordinary freindshipp.[25]

The episode of conflict mentioned here is an unmistakable reference to Tasman's encounter with Māori, which left four of his crew dead. Less obvious, however, is the comment about the 'extraordinary freindshipp' which the English official had to demonstrate to 'procure' this intelligence. This is probably a coded allusion to some form of financial inducement that was necessary in order to obtain these details. Of course, regardless of how much payment the Dutch sources may have received, there was no guarantee of the absolute reliability of the material that they were sharing. One example of this is in the passage describing 'another sort of people about our stature, very white, ruddy, and comely, a people gentle and familiar, with whome, by their owne rellations, they have had some private conference'. It is uncertain which people were being referred to here but, without any corroborating accounts, the English official felt obliged to pass on all the information that he had received without filtering it and leave it for someone else later on to verify.

By the middle of 1644, English EIC officials in London would have received this letter, with its patchy details about New Zealand (although the country was not named in the testimony obtained from the Dutch sources). It is at this point that the flow of information in England about Tasman's expedition apparently comes to an end. No map was reproduced based on the sketch that was sent from Bantam, nor did the partially confused re-telling of Tasman's encounter with Māori make

25. Letter from Bantam, 10 January 1644, India Office Records series IOR/L/MAR/A, British Library, in W. Foster, 'An Early Chart of Tasmania', *The Geographical Journal* 37, no. 5 (1911), 550-51.

its way into any published work or government archives at the time. Instead, this tributary of intelligence dried up – a sure sign that the EIC's directors placed little commercial or even geographic value on it. Who could blame them? Between the struggle to find sufficient subscribers to boost the capital they needed and the ever-present chafing with other European maritime powers in the region, along with supply and staffing problems, and a market that was characterised by its volatility, it is little wonder that EIC officials expressed no zeal for distant territorial exploration – especially in a region which the Dutch had already visited and in which the VOC had evidently found no value.

The particular political circumstances of the English East India Company also militated against any serious enthusiasm for exploration. Unlike its Dutch equivalent, it was not closely entwined with the nation's government. Instead, it tended to act more independently – one of the few advantages of being relatively economically insignificant. Not that the English government was in any position to pay too much attention to the activities of the company anyway, given that it was immersed in what was already proving to be a protracted and ruinous civil war. Hence, the few crumbs of information that the EIC received about New Zealand were swept into a document cupboard somewhere in its head office in Leadenhall Street (which was already 'too cramped to be comfortable')[26] and left to become part of its bureaucratic detritus.

'Staete Landt' Becomes 'Zeelandia Nova'

Already struggling to scrounge a profit from its fickle trading ventures, the English East India Company had no interest whatsoever in Tasman's fleeting visit to some far-off territory – especially one which was clearly of so little value that the Dutch similarly considered the place not being of any commercial consequence. With details of New Zealand's location having ended up filed away and forgotten in London, knowledge of the territory would now enter Europe through the only remaining route: the Netherlands. However, while the Dutch East India Company no longer had any basis for keeping New Zealand's location confidential, as with its English counterpart, neither did it have any reason to promote or circulate details about the territory. It was therefore left to Blaeu – whose role was a curious mixture of employee and entrepreneur – to take the

26. W. Foster, *The East India House: Its History and Associations* (London: John Lane, 1924), p. 16.

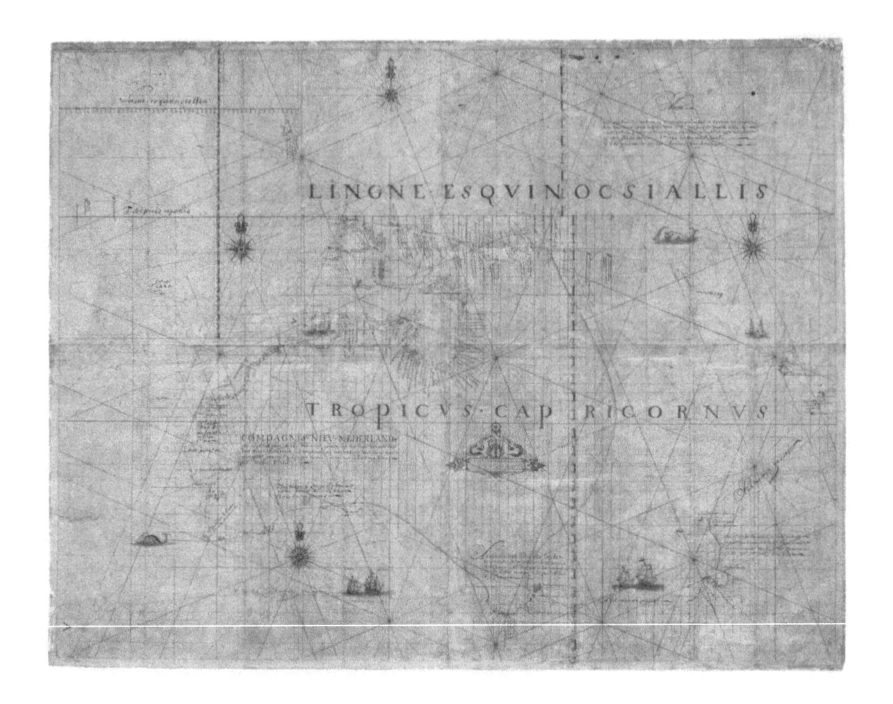

Tasman's kaart van zijn Australische ontdekkingen –
The Bonaparte Tasman Map (1644)
State Library of New South Wales

raw materials of Tasman's expedition and mould them into the sort of cartographic product that would appeal to customers in Europe.

That Blaeu was a master mapmaker is beyond dispute. His almost supreme technical expertise and scientific knowledge were paired with a measured artistic skill, and his works had consequently become the latest word in the graphic rendering of the world's geography in mid-seventeenth-century Europe.[27] However, his urge to profit from his abilities, along with his ambition to be pre-eminent in his profession, led him to take on projects and accept commissions that occasionally were at the limit of his capacity to manage them. At times, his enterprise teetered on the brink of chaos as he attempted to bring some sort of order to the stacks of charts, letters, journals and other documentary materials that littered his office. Working on several projects simultaneously did not help and, unsurprisingly, not all the source material that

27. E. Heawood, 'A Masterpiece of Joan Blaeu', *Geographical Journal* 55, no. 4 (1920), pp. 312-13.

he drew on to create his maps survived the process.[28] Tasman's original manuscript – based on the notes he took on his 1642-43 expedition – is an example of a bundle of papers that was not especially significant in Blaeu's estimation, but that he mined for all he felt might be useful for his next mapmaking venture, after which they were probably put in an out-of-the-way corner of his studio and left in permanent storage.[29] The habit of cataloguing and storing every single document stemming from explorations was one that developed more strongly in the following century and adhered to mainly by governments. Blaeu was running a business, with his reputation and livelihood resting on the accuracy and finesse of the work he produced, rather than on the accumulation of interminable piles of sailors' usually dreary observations for posterity.

However, some records of Tasman's expedition did survive. At least three contemporaneous copies of his journal were made, with one – the so-called 'Hague copy' – signed by Tasman (presumably as a means of authentication). This signed copy is now held in the National Archives in the Hague.[30] Another copy of the journal – almost identical to the Hague copy – is in the British Library and, although it is not signed by Tasman, the person who transcribed it left a note attesting to the fact that it was copied from that signed journal[31] – something that is easily verifiable by the similarities between the texts. Finally, there is the Huijdecoper journal (named after the Huijdecoper van Maarseveen en Nigtevecht family, which was in possession of it from 1844 to 1925).[32] This is housed in the State Library of New South Wales[33] and is also

28. J.C. Stone, 'A Newly Discovered Map of Ettrick Forest, Scotland by Robert Gordon of Straloch: Implications for Sources Consulted by Joan Blaeu', *Imago Mundi* 31, no. 1 (1979), pp. 86-87.
29. B. Hooker, 'Two Sets of Tasman Longitudes in Seventeenth and Eighteenth Century Maps', *Geographical Journal* 156, no. 1 (1990), pp. 26-27.
30. Journaal van Abel Jansz. Tasman van zijn ontdekkingsreis naar het Zuidland, 1642-1643, Zie Aanwinsten, 1.11.01.01, inv. nr. 121 (1867 A III); Tekeningen behorende bij het journaal van Abel Jansz. Tasman van de ontdekking van Van Diemensland en Nieuw-Zeeland, Zie Aanwinsten, 1.11.01.01, inv. nr. 320 (1886 B XV).
31. Copy of the journal of a voyage from Batavia, in the East Indies, for the discovery of the unknown South land, by Abel Jansen Tasman, with sketches of the coast and peoples, British Library, BL Add MS 8946.
32. Burnet, *The Tasman Map*, p. 191.
33. A. Tasman, 'Extract Uittet Journael vanden Scpr Commandr Abel janssen Tasman, bij hem selffs int ontdecken van't onbekende Zuijdlandt gehouden', 1642-1643. Safe 1/72, State Library of New South Wales, ref. 423571.

practically identical to the Hague copy, although without any of the accompanying illustrations (with the exception of two maps). These maps in the Huijdecoper journal have a note attached to them which asserts that they were 'drawn with great diligence and assiduity by Franchoijs Jacobszoon [Visscher], steersman', along with confirmation that in the '[d]raft made by Franchoys Jacobszoon, steersman, the names having been inserted according to directions given by Commander Tasman'.[34] However, there are discrepancies in some of the place-names between the journals, which suggests that the copyists were not as diligent in preparing the maps as they were with the text.

Not only was the original journal that Tasman penned lost later on, but so too were the records from the sister ship on the voyage: the *Zeehaen*. Neither the journal, nor the daily notes made by Visscher have survived, and there is no record of them in any archive. It is known, however, that these documents were copied in Batavia and were among those in the parcel sent to Amsterdam;[35] so, again, circumstances point to them disappearing after Blaeu had finalised his map of the region. The most likely reason for all these journals and maps from Tasman's expedition vanishing is the devastating fire which engulfed Blaeu's entire printing works on 28 February 1672. It was a loss from which the great cartographer never fully recovered (and which may have hastened his death the following year).[36]

* * *

In the mid-1640s, while Blaeu was in his Amsterdam office, assembling and assessing the cartographical information from Tasman's expedition, Tasman was already supervising the production of his own map in Batavia – either in late 1644 or early 1645 – for the benefit of the Dutch East India Company's directors in the Netherlands. Tasman had arrived back in Batavia in August 1644 after his second expedition to Australia, during which time he had carried out further mapping of the continent (the journal accompanying this voyage has also been lost – presumably also in the fire of 1672).[37] It would have made sense to Tasman to consolidate the previous three years' worth of exploration by

34. Cited in Heeres and Coote, *Abel Janszoon Tasman's Journal*, p. 69.
35. R. Meyjes (ed.), *De Reizen van Abel Janszoon Tasman en Franchoys Jacobszoon Visscher Ter Nadere Ontdekking von Het Zuidland in 1642/3 en 1644* (The Hague: Martinus Nijhoff, 1919), pp. 264-68; Heeres and Coote, 61.
36. J.-P. Luminet, 'Willem Janszoon Blaeu' (unpublished paper, Laboratoire Univers et Théories, Paris, 2015), p. 4.
37. R. Gerritsen, 'Getting the Strait Facts Straight', *The Globe* 72 (2013), p. 5.

producing a comprehensive chart of the territories that until that time were unmapped. The result was what is now referred to as the Tasman Bonaparte Map – a hand-drawn work on Japanese paper, measuring 73 by 95 centimetres. The designation Bonaparte stems from the fact that, in 1891, Prince Roland Bonaparte, a grand-nephew of Napoleon and President of the Geographical Society of France, purchased this map. Neither a date nor the name of the cartographer appears on the work, but Gilsemans or Visscher are the most likely contenders for having drafted it.[38]

A possible reason for Tasman organising and overseeing this project is that there was no certainty how long it would take for the cartographers in Amsterdam to complete their own version of the map from the data that had been sent to them from his two expeditions. Tasman may have felt that, in finishing his own map sooner rather than later, it would impress senior company officials over the extent of his accomplishments in these expeditions and stir their interest in his career. Moreover, given the decorative features that appear on this map – such as detailed wind compasses, ships, sea monsters, and various other ornamental flourishes – its purpose was more likely to be as a display piece, perhaps framed and mounted on an office wall in the Company's headquarters, rather than a work produced for navigational purposes. There is uncertainty, however, over when the Tasman Bonaparte Map was published, and although several scholars put the date as late as 1695, they tend to agree that the published version that has survived is a copy of a much earlier work,[39] which has disappeared.

* * *

One piece of information, though, that all the copies of Tasman's journal and maps shared – including the Tasman Bonaparte Map[40] – was the

38. State Library of New South Wales, 'The Tasman Map' (Sydney: State Library of New South Wales, 2020), p. 1.
39. G. Schilder, *Australia Unveiled: The Share of the Dutch Navigators in the Discovery of Australia*, trans. O. Richter (Amsterdam: Theatrum Orbis Terrarum, 1976), p. 354; M. Destombes, *Cartes Hollandaises: La cartographie de la Compagnie des Indes Orientales, 1593-1743* (Saigon: C. Ardin, 1941), pp. 77-79, in J. Tent, 'Who Named New Holland?', Occasional Paper no. 13 (Sydney: Australian National Placenames Survey, 2022), p. 13.
40. Tasman's kaart van zijn Australische ontdekkingen 1644, *de Bonapartekaart*, gereproduceerd op de ware grootte in goud en kleuren naar het origineel in de Mitchell Library, Sydney; met toestemming van de autoriteiten door F.C. Wieder, State Library of New South Wales, ref. MAP NK 1791.

name he gave to New Zealand: 'Staete Landt' (Staten Land). Had he had the time or inclination, Tasman could have circumnavigated the territory and discovered that it was comprised of two main islands. However, having only skirted along a stretch of its west coast, he presumed – somewhat lazily – that the land that he had encountered was part of the vast (and still largely imagined) territory connected to South America, which bore the name Staten Landt.[41] It was not a completely unreasonable deduction, and one that could always be corrected by later explorers. That correction was not long in coming. In 1643, Hendrik Brouwer – who had been instrumental in fixing the preferred sailing route between Amsterdam and Batavia – was on an expedition to Chile, in the course of which he determined that the Staten Landt off the coast of South America was not part of the 'Great Southern Continent'.[42]

When this intelligence reached Amsterdam, shortly after Tasman's journal and maps had arrived, Blaeu had to grapple with how best to deal with the name Tasman had now clearly mistakenly given to the territory lying 1,800 kilometres east of Tasmania. Instead of simply plucking an option from thin air, Blaeu studied some of the other choices that Tasman had made when naming locations in the region, in the hope that he might be able to select one that was in keeping with the sort of thought process Tasman applied to the task. The coast of Western Australia had initially been called Compagnies Nieuw Nederland by Tasman, which he later changed to Hollandia Nova. Given this predilection for using a Dutch placename and qualifying it with 'new', Blaeu repeated this pattern, replacing Staten Land with Zeelandia Nova, after the Netherlands' south-western province of Zeeland. (This later appeared in its anglicised form of New Zealand.[43])

From his workshop overlooking the recently constructed Bloemgracht canal,[44] possibly from as early as 1644, Blaeu began work on a revised map of the world that he planned to publish. He would use the plates from a 1619 world map that his father had made, and etch into them

---

41. Schouten, *Journal ou description du merveilleux voyage de Guillaume Schouten*, p. 24.
42. K.E. Lane, *Pillaging the Empire: Piracy in the Americas, 1500-1750* (New York: M.E. Sharpe, 1998), p. 88.
43. E.H. McCormick, *Tasman and New Zealand: A Bibliographical Study* (Bulletin no. 14) (Wellington: Government Printer, 1959), pp. 10-11.
44. G. Moorman, 'Publishers at the Intersection of Cultures: The Significance of Italo-Dutch Contacts in the Creation Process of Joan Blaeu's *Theatrum Italiae* (1663)', *Incontri* 30, no. 2 (2015), p. 71.

details of Tasman's recent discoveries, including what Blaeu had renamed 'Zeelandia Nova'. This modification of existing maps was a relatively common practice at the time and, in this case, the date of 1619 on the map and the name of the map's original creator – Willem Blaeu – remained on the revised work once it was printed, allowing the younger Blaeu to leverage his father's reputation, as well as save on the considerable expense of producing a completely new set of copper plates.[45]

This was no cottage-industry effort, though. With nine printing presses and a further six used for engravings, Blaeu's printing works were probably the largest in Europe at this time (and had even become an attraction for tourists visiting Amsterdam).[46] Once the copper plates were engraved with all the revised cartographic details and artistic embellishments, the printing presses would crank into action, churning out copies for those customers who were curious about Europe's most recent discoveries. Thereafter, the image of New Zealand (admittedly still in incomplete form) would tip over from being private to public property. The mechanics of mass-produced maps would now expose the farthest frontier of the world (as far as Europe was concerned) to an audience of thousands throughout the continent.

---

45. Tent, 'Who Named New Holland?', pp. 12, 15.
46. A. Briggs and P. Burke, *A Social History of the Media from Gutenberg to the Internet* (Cambridge: Polity Press, 2005), p. 49.

Chapter 5

New Zealand on the Map

Increased Map Publication and Circulation

Maps are illusions. Rather than trying to reproduce physical reality, they offer an allegorical rendition of the terrains that they depict. If anything, the seventeenth-century Dutch efforts to achieve greater cartographical precision went hand in hand with obscuring what the locations being mapped were really like. For a start, all the recently discovered places in the world were named either in Dutch or Latin. To most viewers, this would suggest that those territories were previously nameless, or worse, that the names their indigenous inhabitants had given them somehow did not matter.

Then there were the gaps. In the case of New Zealand, anyone looking at the first published maps of the area would know immediately that the territory was not a single coastline that disappeared into an indifferent expanse of ocean. However, for the next century, these parts of the earth would remain 'without form, and void'[1] to Europeans, and it would be up to future explorers and mapmakers to bring them into being, at least on paper. To a degree, the great illustrated charts of this era took on a form, 'in which European man was the undisputed master of his world', with the cartographer going from outlining the world to insinuating who shaped and controlled it.[2]

However, for all the power vested in cartographers to create an image of the world – at least on paper – there was much that was missing.

1. Genesis 1:2.
2. P. Whitfield, *The Image of the World: 20 Centuries of World Maps* (San Francisco: Pomegranate Artbooks, 1994), p. 74.

In their maps, there was no aesthetic light or shade, no picturesque beauty or sublime terror, no gradients of culture, no variation of language and nothing to signify guilt or redemption. After all, these were ostensibly creations of science, not art. Yet, for all the technical brilliance that they irrefutably possessed, they were just as much works of distortion as they were of veracity. The textures of the world were reduced to two dimensions. Forests, deserts, planes, mountains and populations were noticeable by their absence, and instead of the distinct vibrancy of territories, audiences were given a still-life drained of any vitality. Complexity, colour, intricacy and variation were reduced to monochromatic outlines. The chaos and tangle of the world was tidied up and packed away so that only an abstract order remained. If there was anything to arouse awe in these maps, it was tyranny of distance – the vast expanses of ocean that separated territories from each other – and the matching mystery of the peoples and cultures that inhabited so many of the places that were still barely known to Europe.[3]

By 1645 or 1646, Blaeu had completed his updated version of the 1619 map of the world: *Nova et errariu totius errarium orbis tabula*.[4] However, he was keen to step out of the shadow of his father's reputation and illuminate his own enormous cartographical talents. The recycled 1619 map, which was bordered with 50 images of the (stereo)typical inhabitants of certain lands, and further decorated with stylised classical figures occupying pockets of Arcadia, was already dated by the 1640s and needed to be replaced rather than revised. The steady ascent of the Enlightenment's veneration of science demanded more geographic precision and less superfluous ornamentation in maps. By 1648, Blaeu, who was ever responsive to shifting popular tastes, had developed plans to produce a completely new and larger map (roughly three metres across and two metres high). It would show the world in two hemispheres,[5] appearing side by side (a further abstraction from the reality of the

3. A. Zukas, 'Class, Imperial Space, and Allegorical Figures of the Continents on Early-Modern World Maps', *Environment, Space, Place* 10, no. 2 (2018), 29-62.
4. Tent, 'Who Named New Holland?', pp. 12, 15; Schilder, 'New Cartographical Contributions', p. 42; Schilder, *Australia Unveiled*, pp. 364, 366, 370; K. Zandvliet, 'Golden Opportunities in Geopolitics: Cartography and the Dutch East India Company during the Lifetime of Abel Tasman', in W. Eisler and B. Smith (eds), *Terra Australis: The Furthest Shore* (Sydney: International Cultural Corporation of Australia, 1988), p. 80; W. Blaeu and J. Blaeu, *Nova et accurata totius terrarum orbis tabula* (1645-46 [1619]), Collecties Maritiem Museum 'Prins Hendrik', Rotterdam, ref. K259.
5. Meyjes (ed.), *De Reizen van Abel Janszoon Tasman*, p. 267.

world as a sphere), and would be designed in a more modern and more commercially appealing style.

Blaeu envisaged that this grand and expensive work would make an imposing statement when hung on the wall of a wealthy merchant or senior government official, and therefore fashioned it specifically for display, rather than for storage in a collection. Of course, such a sumptuously designed artefact would also exalt Blaeu's cartographical mastery[6] and enhance his firm's reputation, which was yet another reason for manufacturing this new map. The title he gave to this work was *Nova totius errarium orbis tabula* but it has since more commonly come to be known as the Blaeu World Map.[7] It was based on a series of new copperplate engravings, which reproduced the map in black and white as the first stage in the production process. Each map was then individually hand-coloured and supplemented with a text beneath the images, giving some description of their content.[8]

The ongoing spread of literacy throughout Europe in this era, accompanied by the growth of a more wealthy middle class,[9] had created an expanding market of people who were both keen to inhale an atmosphere of academic curiosity and able to pay for the privilege. In the deeply Protestant Netherlands at this time, though, overt displays of material abundance were generally frowned upon. Excessive affluence was seen as the precursor of moral decay. However, as prosperity grew to new heights in cities such as Amsterdam, the tension between restraint and extravagance became more pronounced.[10] In their small way, maps were

6. D. van Nette, 'The New World Map and the Old: The Moving Narrative of Joan Blaeu's Nova Totius Terrarum Orbis Tabula (1648)', in B. Vannieuwenhuyze and Z. Segal (eds), *Motion in Maps, Maps in Motion: Mapping Stories and Movement through Time* (Amsterdam: Amsterdam University Press, 2020), pp. 1-6.
7. Blaeu dedicated it to the Spanish ambassador, Gaspar de Bracamonte Guzmán, count of Peñaranda de Bracamonte. M. Debergh, 'A Comparative Study of Two Dutch Maps, Preserved in the Tokyo National Museum: Joan Blaeu's Wall Map of the World in Two Hemispheres, 1648, and Its Revision *ca.* 1678 by N. Visscher', *Imago Mundi* 35, no. 1 (1983), p. 20.
8. J. Blaeu, *Nova totius terrarum orbis tabula* (1648), Harry Ransom Center, University of Texas at Austin, Map_Kraus_03; Debergh, 'A Comparative Study', p. 22.
9. R. Houston, 'Literacy and Society in the West, 1500-1850', *Social History* 8, no. 3 (1983), 269-93.
10. S. Schama, *The Embarrassment of Riches: An Interpretation of Dutch Culture in the Golden Age* (New York: Vintage Books, 1997), pp. 47-55.

a means of overcoming the dilemma of how to parade wealth modestly. A large, new, hand-painted map was expensive but could be morally justified because it would hang on a wall without being too conspicuous, with the added advantage of serving an educative function and being emblematic of the blessings that the Dutch Empire had received.

* * *

Not all buyers, though, would necessarily have had the space to display a copy of *Nova totius errarium orbis tabula* and so, to take further advantage of this burgeoning market for geographical knowledge, Blaeu also manufactured globes containing the same details that appeared on his wall maps, but on a smaller scale. The exact number he produced is unknown but, in 1976, there were still 86 of the 1648 version of this globe known to be in existence.[11] Blaeu's globe featuring New Zealand probably preceded his wall map by a few months. This likelihood arises from the discovery in Copenhagen of one of the cartographer's 1648 globes, which bore a dedication to King Christian IV of Denmark, who had died on 28 February that year,[12] while Blaeu's wall maps were dedicated to the ratification of the Treaty of Münster on 15 May 1648.[13]

Thus, within just a few years of Tasman reaching New Zealand and partially surveying its coast, the information he had pieced together had gone back to Batavia, been examined and evaluated by officials there, then copied and sent to Amsterdam, where it was once again assessed (this time by the directors of the Dutch East India Company). From there, it was passed on to the VOC's chief mapmaker, who converted it first, around 1645, into an update of an earlier map and then, three years later, into a work of cartographical art – all with the tacit blessing of the company and the Dutch government. This route had taken the fragmentary outline of New Zealand from a Dutch discovery to a European commodity – an amalgam of intelligence that could be purchased on a single sheet. What was left behind in this race to meet customer demand, however, was any systematic means of subsequently updating

11. T. Campbell, 'A Descriptive Census of Willem Blaeu's Sixty-Eight Centimetre Globes', *Imago Mundi* 28, no. 1 (1976), p. 36.
12. Ibid., p. 30; G. Schilder, 'Die Entdecking Australiens im Niederländischen Globusbild des 17 Jahrhunderts', *Der Globusfreund* 25 (1978), p. 189.
13. F.C. Wieder, *Monumenta Cartographica: Reproductions of Unique and Rare Maps*, 3 (The Hague: Martinus Nijhoff, 1925-33), p. 61; L.M. Baena, 'Negotiating Sovereignty: The Peace Treaty of Münster, 1648', *History of Political Thought* 28, no. 4 (2007), p. 618.

these maps. Many map publishers only had passing contact with the company (if at all) but, even if they were committed enough to scour its correspondence for snippets of information to enhance their maps (presuming that they could get the necessary permission in the first place), for most of them it was both faster and cheaper simply to replicate rather than research.

Certainly, not all of those involved in the mapmaking business approached their work as assiduously as Blaeu. At the same time that he was putting the finishing touches on his 1648 world map, just a kilometre's walk eastwards, a marble mosaic of a two-hemisphere map of the world was being pieced together for the floor area inside the entrance of the Amsterdam town hall. The city's civic and imperial activities had not only become inseparable but had now visually merged. However, in what could be interpreted as a metaphorical warning of the fleeting nature of empires, the mosaic began to wear under the constant foot traffic of Amsterdam's citizens visiting the building and had to be replaced in 1746 with a more conventional floor.[14] Eighty-five years before its removal, an image of this mosaic map – *De Grondt en Vloer vande Groote Burger Sael* – was produced by the engraver Jacob Vennekool. Oddly, the outline of New Zealand's coast was missing (even though it had already appeared in Blaeu's maps by this time). Blaeu may have had a hand in this project[15] but nothing featured between the east coast of Australia and South America appeared.[16] Thus, *De Grondt en Vloer vande Groote Burger Sael* was one of the last maps of the globe produced without New Zealand appearing on it.

* * *

Once Blaeu's world map and globe were in circulation, imitators were certain to surface. Publishing was a major industry in the Netherlands, with around 350,000 works printed in the republic in the seventeenth century.[17] Reproductions could be made by printers who did not need

---

14. G. Schilder, *Monumenta Cartographica Neerlandica*, 9 vols (Alphen aan den Rijn: Canaletto, 2000), Vol. 6, p. 7.
15. E.-J. Goossens, *Treasure Wrought by Chisel and Brush: The Town Hall of Amsterdam in the Golden Age* (Zwolle: Waanders, 1996), p. 28; Sharp, *The Voyages of Abel Janszoon Tasman*, p. 342.
16. J. Vennekool, *De Grondt en Vloer vande Groote Burger Sael: Tot Amsterdam: Bij Dancker Danckerts* (1661), State Library of New South Wales, M2 100/1661/1.
17. A. Pettegree and A.T. der Weduwen, 'What Was Published in the Seventeenth-Century Dutch Republic?', *Livre. Revue Historique* (2018), 1-27.

to incur the expense of researching and designing maps and, therefore, could be sold at a lower price than the original works on which they were based. A rudimentary form of temporary copyright did exist, with the Dutch authorities able to issue what were known as 'printing privileges', but obtaining these was not straightforward and they proved difficult to enforce, resulting in some printers never bothering to apply for them at all.[18] Blaeu's 1648 map contained an inscription asserting privilege and protection from errariumd copying but, within a decade, it was evident that such a measure to protect his intellectual property was completely ineffectual.[19]

New Zealand now appeared on the horizon of European knowledge and, with every subsequent map that was published in the seventeenth century, it became a firmer fixture in Europe's conception of the world's form. A complete inventory of all the maps depicting New Zealand in the second half of the century is now practically impossible to assemble, as it cannot be known for certain precisely how many were produced, or by whom. However, there are sufficient examples to illustrate how the progeny of Blaeu's world map proliferated in the decades following its publication.

One of the first published maps which clearly borrowed from Blaeu's great work appeared in 1652. *Terra Australis Incognita*[20] was the work of Jan Janszoon, an Amsterdam cartographer and printer who had married Elisabeth de Hondt, the daughter of one of Blaeu's main competitors, Jodocus de Hondt.[21] By the 1640s, Janszoon had entered the pantheon of great Amsterdam map publishers and was a budding rival to the dominance of the Blaeu dynasty.[22] *Terra Australis Incognita* was one of his more modest works, though, encompassing the Southern Hemisphere area from the South Pole to roughly 30 degrees longitude

18. M. Buning, 'Privileging the Common Good: The Moral Economy of Printing Privileges in the Seventeenth-Century Dutch Republic', in S. Graheli (ed.), *Buying and Selling: The Business of Books in Early Modern Europe* (Leiden: Brill, 2019), pp. 88-89.
19. Hooker, 'New Light on the Mapping and Naming of New Zealand, p. 162.
20. J. Janszoon, *Terra Australis Incognita* (1652), Harvard Library, Cambridge MA, Harvard Map Collection, ref. MAP-LC G9801.S12 1652. J3.
21. National Library of Australia, *Mapping Our World: Terra Incognita to Australia* (Canberra: National Library of Australia, 2013), p. 157.
22. J. Lydon, 'Visions of Disaster in the Unlucky Voyage of the Ship *Batavia*, 1647', *Itinerario* 42, no. 3 (2018), p. 353; L. Isdale, 'Janszoon in Context: Cartography from Earliest Times to Flinders', *Queensland History Journal* 21, no. 2 (2010), p. 111.

and measuring 44 by 49 centimetres. 'Nova Zeelandia' appears on the work, accompanied by mention of the fact that Tasman had discovered the territory in 1642 – information no doubt obtained directly from Blaeu's map. The margins of Janszoon's work were decorated with samples of generic indigenous people, whose freedom from the burdens of civilisation – at least as far as European thought was concerned in this era – deprived them of full property rights over their territories (the principle of *terra nullius*).[23] They are pictured in poses involving cooking, dancing or just loitering. This was not a new map by Janszoon, however, but simply a slightly revised version of *Polus Antarcticus*, which had first been published by the Dutch cartographer and engraver, Henricus Hondius – Janszoon's brother-in-law – around 1637 (and which at that time obviously had New Zealand missing).[24]

* * *

One of the greatest publishing ventures by volume in the Netherlands during the seventeenth century commenced in 1637, when the Dutch government errarium the printing of a version of the Bible in the country's own language. Known commonly as the Ravesteyn Bible (after its publisher), it became the equivalent of the King James Bible in England – part of the central effort during the Reformation to ensure that the Scriptures were available to people in their own vernacular, rather than being locked into Latin as had been the case for centuries in Europe.[25] In the following decades, new editions regularly flowed from the presses to meet a seemingly insatiable popular demand among Dutch Protestants. Some of the more expensive versions came with illustrations, including a 1657 edition containing a world map created by the cartographer Nicolaus Joannes Visscher (who also published this edition of the Bible).[26] This was known as the *Orbis errarium nova et*

23. M. Borch, 'Rethinking the Origins of Terra Nullius', *Australian Historical Studies* 32, no. 117 (2001), 222-39; B. Charry, *The Tempest: Language and Writing* (London: Bloomsbury, 2013), p. 107.
24. H. Hondius, *Polus Antarcticus* (Amsterdam, 1637), State Library of Western Australia, Globe, no. 54, 2003 0311-3930, CIU file ref. 16/212.
25. H. Nellen and P. Steenbakkers, 'Biblical Philology in the Long Seventeenth Century: New Orientations', in D. van Miert, H. Nellen, P. Steenbakkers and J. Touber (eds) *Scriptural Authority and Biblical Criticism in the Dutch Golden Age: God's Word Questioned* (Oxford: Oxford University Press, 2017), p. 46.
26. E. Stronks, 'The Diffusion of Illustrated Religious Texts and Ideological Restraints', in A. Lardinois, S. Levie, H. Hoeken and C. Lüthy (eds), *Texts,*

accuratissima tabula[27] and was yet another variant of Blaeu's 1648 map. Ironically, despite the commitment by these Protestant printers and publishers to consign Latin to the past, when it came to map titles, Latin continued to be used, as it was still regarded as the language of science.

The significance of Visscher's work lay not so much in its cartographical value, but in the fact that it became the first map containing an image of New Zealand that was truly mass-produced, with thousands of copies in circulation in the Bibles that he published. In one version of this work, the space around the two hemispheres in the map contained allegorical representations of the four continents: Europa – a sort of Britannia for the continent – dominated the world; Asia (seated on a camel) was dressed in what would have been popularly imagined as eastern clothes and swinging a censer; Africa was perched on what looks vaguely like a crocodile; and America – clutching an axe and bow, was astride an armadillo.[28] This sort of imagery – as jingoistic and crude as it was – stimulated the imagination of readers when looking at what would otherwise be simply the outlines of territories. It provided a framework for interpreting the physical and social geography of the world.

Not wishing to be overtaken by these pretenders, Blaeu issued a revised version of his world map in 1659. Accompanying the map was an explanatory note, part of which read:

> But of all these and of the above-mentioned islands we cannot speak more fully because of the want of space; nor has there yet been published anything, or little concerning these last named; wherefor the reader and spectator must rest content with this map, until I, Blaeu, shall publish these and the aforesaid a large book, full of maps and descriptions, which is at present being prepared.[29]

Transmissions, Receptions: Modern Approaches to Narratives (Leiden: Brill, 2015), p. 213.

27. *Orbis terrarum typus de integro in plurimis emendatus, auctus, et icunculis illustratus*, Sanders of Oxford, Antique Prints and Maps, stock I.D. 43111.
28. Ibid.
29. J. Blaeu, *Archipelagus Orientalis sive Asiaticus* (Amsterdam: Joan Blaeu, 1659), paper on linen with wooden rollers. Large wall map of Southeast Asia and Australia in six engraved map-sheets, three panels of letterpress descriptive text in Latin, Dutch and French with imprint of Blaeu in the three languages, inset maps of Borin and the Solomon Islands, contemporary hand colour, backed on original linen, original decorated rollers, total dimensions 158.7 × 117.4cm, Sotheby's lot 93 (May 2017).

There was no point going to other publishers, Blaeu was telling his readers: Wait until I have new information to hand which I shall release shortly. However, the planned magnum opus was never published, possibly because of the enormous time it would have taken to produce and the corresponding costs of publishing it in a market where maps of the world were becoming commonplace. Blaeu's intentions do reveal, though, something of the highly competitive atmosphere within which Dutch cartographers were operating, as well as his personal ambition to remain known as the pre-eminent mapmaker in the country.

It was as though the entire printing industry in Amsterdam was looking to profit from producing maps and, by the 1660s, works were appearing that were copies of copies of original maps. Visscher's 1657 world map – which was a near-duplication of Blaeu's – was itself copied three years later, probably by Justus Danckerts, who belonged to another prominent family of cartographers and artists. The geographical information in Danckerts' *Nova totius errarium orbis tabula* was nearly identical to the maps he was emulating. Only the surrounding ornamentation differed. Gone were the continental allegories and in their place were lavish scenes from classical mythology.

Not surprisingly, precision was sometimes sacrificed by the time that maps were appearing that were third- or fourth-hand copies. This can be easily seen in Frederick de Wit's 1660 map *Nova totius errarium orbis tabula*,[30] in which Tasman's outline of New Zealand's west coast was depicted almost as a straight line. An equally distorted rendition of New Zealand appeared in *Pascaerte Vande Zuyd-Zee Tussche California, en Ilhas de Ladrones*,[31] a map published in 1666 by the cartographer Pieter Goos, who had previously worked with Janszoon and Hondius. Instead of a straight line, though, as had appeared in de Wit's map, Goos portrayed the territory as being dominated by one huge bay, which had a coast that ran for the same distance as the length of the South Island.

Another clue revealing the casual approach some printers and publishers took when replicating existing maps can be found in Johannes van Keulen's 1680 work, *Pascaert vande Zuyd Zee en een gedeelte van*

---

30. F. de Wit, *Nova totius terrarum orbis tabula* (1660), Library of Congress Geography and Map Division, Washington, DC, G3200 1660.W5, Library of Congress Control Number 2006627253.
31. P. Goos, *Pascaerte Vande Zuyd-Zee tussche California, en Ilhas de Ladrones* (1666), Yale University Library, Beinecke Rare Book and Manuscript Library, Call Number 23cea 1666, Container/Volume BRBL_00682.

Brasil van ilhas de Ladronos tot R. de la Plata.[32] The tell-tale sign that van Keulen took little care when reproducing his maps from the latest samples can be found in the fact that the name he used for New Zealand was Staten Landt, even though this designation had ceased to appear on most maps a quarter of a century earlier.

It would be easy to lay the blame for such lapses on the more modest abilities of some of the cartographers involved in publishing these maps. Yet, in 1672, when Blaeu completed his multi-volume *Atlas Maior*,[33] there was another error in how New Zealand appeared. *Atlas Maior* was the culmination of Blaeu's career. At 3,368 pages in length and containing 594 maps, there was no other work this substantial (or expensive) anywhere in Europe at this time, and subsequent editions were published in French, Dutch, German and Spanish (although, oddly, not in English). The problem with the depiction of New Zealand in the world map featured in this monumental work was that it was not taken from his 1648 double-hemisphere map but, for some unknown reason, from a copy made by one of his competitors.[34] Just by chance, however, Blaeu's error in this great atlas was to render New Zealand as two islands, separated by what would later become known as Cook Strait. This diversion from Tasman's chart, the soul authoritative source at Blaeu's disposal, turned out to be the first time New Zealand was mapped this way, but it was the product of lethargy rather than strategy.

Further maps featuring New Zealand emerged in the Netherlands in the closing decades of the seventeenth century, almost all of which were variations on the basic theme produced by Blaeu in the mid 1640s. The early period of experimentation in mapping the South Pacific had given way to an era of repetition and, in the course of this transition, New Zealand had gone from being a minor novelty on world maps to something of an inconsequential outlier. Variations of that small geographical curvature printed in a remote part of the remote hemisphere promised little to those who even noticed it. However, it was

32. J. van Keulen, *Pascaert van de Zuyd Zee en een gedeelte van Brasil van ilhas de Ladronos tot R. de la Plata* (1680), Beinecke Rare Book and Manuscript Library, Yale University Library, Call Number 23cea 1697, Container/Volume BRBL_00682.

33. Sometimes known as *Cosmographia Blaviana*.

34. P. van der Krogt, *Koeman's Atlantes Neerlandici New Edition: Volume 2: The Folio Atlases Published by Willem Jansz. Blaeu and Joan Blaeu* (Utrecht: Hes & De Graaf, 2000), p. 485.

now attached to the body of Europe's cartographical knowledge and, wherever maps of the world were published from now on, New Zealand was certain to feature on them.

The Establishment of the Royal Society in England

By the early 1660s, the great English Restoration diarist, Samuel Pepys, was thoroughly enjoying his financially comfortable career on the Navy Board, along with a bustling and often indulgent social life.[35] As well as chronicling the great events that marked out the capital's history – such as the plague of 1665 and the Great Fire of London the following year – Pepys also documented many of his more mundane daily activities. One such event took place on 8 September 1663. 'Dined at home with my wife', he wrote contentedly in his journal that evening:

> we had a good pie baked of a leg of mutton; and then to my office, and then abroad, and among other places to Moxon's, and there bought a payre of globes cost me L3 10s., with which I am well pleased, I buying them principally for my wife, who has a mind to understand them, and I shall take pleasure to teach her. But here I saw his great window in his dining room, where there is the two Terrestrial Hemispheres, so painted as I never saw in my life, and nobly done and to good purpose, done by his own hand.[36]

The shop referred to in this entry was owned by Joseph Moxon and was easily spotted at its location by its prominent shop sign, which pictured the mythical figure of Atlas shouldering the world (which became a trademark for his business).[37] Moxon created, printed and sold his own maps, globes and books, along with mathematical equipment, and it was at this shop – in Cornhill, close to Leadenhall Market – that the first map which depicted New Zealand was printed and published in England in 1657.

35. R. Latham (ed.), *The Shorter Pepys* (London: Bell & Hyman, 1985), pp. xxv-xxvi.
36. S. Pepys, *The Diary of Samuel Pepys, MA, FRS, 1663* (Frankfurt: Outlook, 2018), p. 191.
37. He even mentioned his 'At the Sign of the Atlas' on the publication page of his own books.

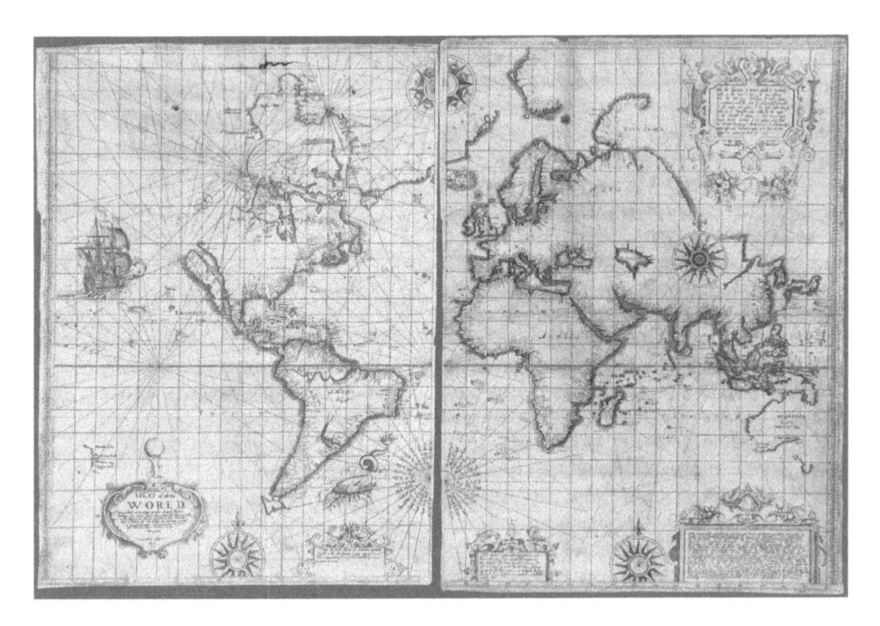

A plat of all the world (Joseph Moxon, 1657)
Boston Public Library, Mapping Boston Collection

This migration of the world map to England was not the result of a fleeting encounter with Dutch cartography, but the result of a lengthy, albeit uneven, apprenticeship that Moxon had served with mapmakers and printers in the Netherlands. His father – a Puritan – had moved to Delft in 1637 to escape the increasing persecution being meted out to non-Anglicans in England at this time. There, he began work printing religious material and, during this time, his son Joseph became fluent in Dutch and familiar with printing industry in the country. Both father and son returned to England in 1643 but, nine years later, Joseph travelled back to Amsterdam, where he had an encounter that piqued his curiosity in the discovery of new nautical passages:

> I went into a Drinking-house to drink a cup of Beer for my thirst, and sitting by the publick Fire, among several People there hapned a Seaman to come in, who seeing a Friend of his there, who he knew went in the *Greenland* Voyage, wondred to see him, because it was not yet time for the *Greenland* Fleet to come home, and ask'd him what accident brought him home so soon: His Friend (who was the Steerman aforesaid in

> a *Greenland* Ship that Summer) told him that their Ship went not out to Fish that Summer, but only to take in the Lading of the whole Fleet....I entred discourse with him, and seem'd to question the truth of what he said. But he did ensure me it was true, and that the Ship was then in *Amsterdam*, and many of the Seamen belonging to her to justifie the truth of it.[38]

Over a chat with a stranger in a pub, Moxon learned about the north-east polar passage to Japan and added it to his store of geographical knowledge as he developed his globe- and mapmaking skills during this time.[39] (In yet another Dutch connection, when Moxon published a book on navigation in 1657, he dedicated it to Thomas Whetstone, who was born in the Netherlands and whose uncle was Oliver Cromwell.)

By 1654, Moxon was back in London, writing mathematical books and publishing the works of several English scientists. Gradually, his own scientific expertise and reputation grew to the point where, in 1662, he was appointed Royal Hydrographer to Charles II.[40] Cartography had been a relatively minor part of Moxon's activities upto this time and he engaged in it as much for commercial purposes as from scientific curiosity, as he conceded when he published one work on navigation, which he was candid enough to dedicate 'to the Profit of the Printer, the benefit of the Buyer, and the general profit and benefit of the whole Art of Navigation'.[41]

On his map, *A plat of all the world*,[42] which was part of an atlas that he published in 1655,[43] Moxon described how it was:

> Projected according to the truest Rules Being far more exact than either the Plain-Card or the Maps of the World described

38. J. Moxon, *A Brief Discourse of a Passage by the North Pole to Japan, China. &c* (London: Joseph Moxon, 1674), pp. 1-2.
39. G. Jagger, 'Joseph Moxon, FRS, and the Royal Society', *Notes and Records of the Royal Society of London* 49, no. 2 (1995), p. 194.
40. Ibid., pp. 194-97.
41. E. Wright and J. Moxon, *Certaine Errors in Navigation Detected and Corrected, with Many Additions that were not in the Former Editions* (London: Joseph Moxon, 1657), n.p.
42. J. Moxon, *A plat of all the world: projected according to the truest rules being far more exact than either the plain-card or the maps of the world described in two rounds*, Boston Public Library, Mapping Boston Collection, Identifier: 06_01_000090, Barcode: 39999052509518.
43. G.D. Gitzen, 'Edward Wright's World Chart of 1599', *Terrae Incognitae* 46, no. 1 (2014), p. 7.

> in two Rounds First set forth by Mr Edw. Wright and now newly Corrected and inlarged with many New Discoveries by Jos. Moxon And Sold at his Shop in Cornhill at the Sign of Atlas.[44]

The 'Edw. Wright' he mentioned on the map was the mathematician and cartographer Edward Wright,[45] who in 1599 had published *Certaine Errors in Navigation*,[46] after a voyage to the Azores. Moxon eventually revised this work and reprinted it in 1657,[47] including his new world map as one of the features of it.

New Zealand was one of the territories added to Wright's map by Moxon. It was referred to as Zelandia Nova, with its shape looking remarkably similar to that of Goos' *Pascaerte Vande Zuyd-Zee tussche California, en Ilhas de Ladrones*, which appeared nine years later (raising the strong possibility that Moxon's work may have pollinated some of the details in subsequent Dutch maps). However, Moxon appended only three names to New Zealand's coast and used a combination of Dutch and English words ('*hoeck*' and 'bay', respectively) when labelling certain locations.

Design-wise, Moxon's world map was immediately distinguishable from its European counterparts. While the Dutch maps of this time were almost becoming an over-ornamented parody of themselves – with their embellishments having gone from creative to cliché – Moxon's 1655 work had most of the filigree stripped away. The large, royal coat of arms in the top left-hand corner had been erased from the original plate, a new cartouche was engraved in the lower left-hand corner,[48] and two sea-creatures and a ship were the only (faint) decorative concessions that remained. He saw himself as a serious technician rather than an artist and, accordingly, invested his attention in measurements and coordinates.

The vague intelligence about New Zealand's existence and location that had reached London from an outer branch of the English East India Company thirteen years earlier had withered and disappeared

44. Moxon, *A plat of all the world.*
45. E.J.S. Parsons and W.F. Morris, 'Edward Wright and His Work', *Imago Mundi* 3, no. 1 (1939), pp. 68-69.
46. E. Wright, *Certaine Errors in Navigation*, (London: Valentine Sims, 1599).
47. Wright and Moxon, *Certaine Errors in Navigation Detected.*
48. Pierre S. DuPont III Collection of Navigation, auction, Christie's New York, 8 October 1991, lot 254.

into an archival crevice in the company's headquarters. (This building, coincidentally, was located just 150 metres away from Moxon's shop.) Now, though, it was sprouting afresh in a new map that was being produced commercially. No records of the sales of Moxon's 1655 atlas or his 1657 edition of *Certaine Errors* exist, but there are intriguing possibilities about the extent to which his world map began to infiltrate the English imagination.

One immediate documented example comes from Pepys himself. His diary entry from the autumn of 1663 records that he purchased a pair of globes from Moxon's shop, partly in order to share with his wife his own particular interest in geography. In ways such as this, an awareness of the territories of the world known to Europe at this time were able to circulate easily among those with an interest in the form of the world. More speculatively, Pepys' job as an official on the Navy Board[49] meant that such information – new as it was to most English readers – was starting to contribute to the architecture of knowledge about the world that was being assembled in official British circles. While there was no official appetite at this time for exploring the distant region where New Zealand lay, such remote and 'unclaimed' territories were nonetheless sources of latent curiosity.

The fact that Moxon's 1657 book and accompanying world map appeared in English was a crucial factor in popularising knowledge about New Zealand's existence and location among the British. As the case of Tasman's map had exemplified, knowledge in so many science-related disciplines travelled swiftly throughout Europe and was translated into several of the continent's languages before belatedly appearing in English versions (if at all). The Welsh mathematician Robert Recorde drew attention to this situation in the previous century. 'Sore oftentimes have I lamented with my self', he wrote in the preface to his book on arithmetic, 'the infortunate condition of England, seeing so many great clerks to arise in sundry other parts of the world, and so few to appear in this our nation: whereas for excellencie of natural wit (I think) few nations do match English men.'[50] By the mid-seventeenth century, though, England's intellectual insularity was giving way to scholars, publishers and general readers absorbing all that their European neighbours had to offer and, more than that, they were adding to this knowledge, both by refining what was being produced in other countries and, gradually,

49. R. Latham (ed.), *Samuel Pepys and the Second Dutch War: Pepys's Navy White Book and Brooke House Papers* (Oxford: Routledge, 2019), p. i.

50. R. Recorde, in K. James, 'Reading Numbers in Early Modern England', *BSHM Bulletin* 26, no. 1 (2011), p. 5.

when it came to cartography, through their own fledgling efforts at maritime exploration. The result was a profusion of English-language texts tackling all manner of scientific subjects.[51]

Moxon himself – piloted by ambition – moved increasingly in the direction of becoming a gentleman scholar, which took him and his work into a more rarefied echelon. He got to know contemporary mathematical luminaries such as Isaac Newton, Walter Pope, Elias Ashmole and Jonas Moore and, when he sought the appointment as Royal Hydrographer, eleven of his thirteen nominees were Fellows of the Royal Society.[52] His connection with the Royal Society extended further his web of professional associates and many of the society's members, for their part, evidently held Moxon in high regard. In 1664, the society's council agreed to lend him some of its scientific equipment and also approved the purchase one of his globes. Moreover, when it came to deciding who would be sent to Moxon's shop to buy the globe, the council delegated that responsibility to the greatest British architect of the age, Sir Christopher Wren[53] (to whom Moxon later dedicated one of his books[54]). In the following decade, Moxon also developed a close professional and personal association with Robert Hooke,[55] who from 1665 was Curator of Experiments at the Royal Society.[56]

* * *

The Royal Society was established in 1660 and received a Royal Charter two years later. In the opening paragraph of the Charter, Charles II outlined how the society's purposes fitted in with those of the nation.

---

51. Ibid., p. 7; K. Raine, 'Blake's "Marriage of Heaven and Hell" by Martin K. Nurmi', *The Modern Language Review* 53, no. 2 (1958), p. 246.
52. Jagger, 'Joseph Moxon, FRS, and the Royal Society', pp. 194-98; D.A. Long, *'At the Sign of Atlas': The Life and Work of Joseph Moxon, A Restoration Polymath* (Donington: Shaun Tyas, 2013), pp. 25-26, in J. Hargrave, 'Joseph Moxon: A Re-fashioned Appraisal', *Script & Print* 39, no. 3 (2015), p. 171
53. Jagger, 'Joseph Moxon, FRS, and the Royal Society', p. 194; The Royal Society, Repository GB 117, ref. CMO/1/46 (30 March 1664), pp. 61-62.
54. J. Moxon, *Regulae Trium Ordinum Literarum Typographicarum, or, The Rules of the Three Orders of Print Letters* (London: Joseph Moxon, 1676), p. ii.
55. Hargrave, 'Joseph Moxon: A Re-fashioned Appraisal', p. 171; R. Iliffe, 'Material Doubts: Hooke, Artisan Culture and the Exchange of Information in 1670s London', *British Journal for the History of Science* 28, no. 3 (1995), p. 307; F. Henderson, 'Robert Hooke and the Visual World of the Early Royal Society', *Perspectives on Science* 27, no. 3 (2019), 395-434.
56. R. Waller, *The Posthumous Works of Dr. Robert Hooke* (London: Sam Smith, 1705), p. i.

'We have long and fully resolved', the king asserted, 'to extend not only the boundaries of the Empire but also the very arts and sciences.' The marriage of science and the state had now been consummated, with the promise of imperial progeny not far off. 'We look with favour upon all forms of learning', the king continued but, perhaps inspired by the university systems in Europe, he was encouraging of those branches 'which by actual experiments attempt either to shape out a new philosophy or to perfect the old', in the hope that 'such studies, which have not hitherto been sufficiently brilliant in any part of the world, may shine conspicuously amongst Our people'.[57] Modern science was still in its infancy at this time but, already, the small steps it was taking were being anticipated by at least a few people as a portent of great imperial strides at some point in the future.

Charles II's interest in the work of the Royal Society – particularly the part it might play in Britain's future imperial ambitions – could well have been awoken by a gift he was given on ascending the throne in 1660. It was an atlas presented to him by a delegation of Dutch traders, led by an Amsterdam-based academic, Johannes Klencke. The atlas, entitled *Soli Britannico Reduci Carolo Secundo regum augustissimo hoc Orbis Terrae Compendium*, was one of several items given to Charles II in the hope that it would go some way towards encouraging improved trading relations between the two nations,[58] and comprised of a collection of 42 maps of various parts of the world, including a chart of Australia and New Zealand, based on Blaeu's world map.[59] The atlas was modestly decorated and hand-coloured (perhaps anticipating British tastes at this time) and came with detailed written information for each of the mapped areas in the work.[60] It was the sort of publication that the king

57. Royal Society Charter, 1662, in M.A. Peters and T. Besley, 'The Royal Society, the Making of "Science" and the Social History of Truth', *Educational Philosophy and Theory* 51, no. 3 (2019), p. 227.
58. P. Barber, *The Map Book* (London: Weidenfeld & Nicolson, 2005), p. 164.
59. J. Klencke (ed.) *Soli Britannico Reduci Carolo Secundo regum augustissimo hoc Orbis Terrae Compendium* (Amsterdam: Visscher, Blaeu and others, 1660).
60. *Soli Britannico Reduci Carolo Secundo regum augustissimo hoc Orbis Terrae Compendium humill. off. I. Klencke* [A collection of forty-two maps of all parts of the world, published by J. and W. Blaeu, H. Allard, N.J. Visscher and others, made up into a volume by J. Klencke and other merchants of Amsterdam and presented by them to King Charles II of England at his accession in 1660.] 1613-1660', British Library, System number 004959010, UIN: BLL01004959010, Shelfmark: maps K.A.R.

could leaf through and obtain a good sense of the latest details on the world's territories, as depicted by Europe's greatest cartographers.

Charles II therefore had good reason to encourage home-grown research into the geography and cultures of the world, aware of the possible scientific and political gains to be made. At this stage, though, the Royal Society still had the outward appearance of a gentlemen's club, rather than a serious academic institution, with some regarding its scientific pretensions as little more than eccentricities.[61] However, in the second half of the seventeenth century, Britain was on the verge of becoming a major imperial power and, as its economic, cultural and scientific dominion began to grow, the society would serve as one of its intellectual engine rooms. Part of this imperial expansion was only possible because of the way in which usually random elements of intelligence were assembled together for what was a broadly singular purpose. For the remainder of this century (and for most of the following one), the Royal Society, politicians and officials, academics, writers, publishers, mariners and merchants in Britain would contribute pieces to the puzzle and, gradually, the image of the Empire took shape. Maps were initially just one small part of this machinery of expansion, but their importance quickly magnified as the nation began to mature as a maritime power. For the moment, though, there was no urgency to acquire additional cartographical information for imperial purposes, and so those in Britain (most of them still hobbyists) with an interest in this field were content to open their atlases in their libraries or reading rooms and examine them more as intriguing artefacts of geographical knowledge rather than the embodiments of egregious triumphalism that they later became.

61. S. Schama, *A History of Britain: Volume 2: The British Wars 1603-1776* (London: The Bodley Head, 2009), p. 215.

Chapter 6

The Growth of Literacy and Mapmaking in England

The First English Map of the South Pacific

For the remainder of the seventeenth century, the partial map of New Zealand remained stagnant in England. After all, there was nothing more that could be added to what Moxon had published in 1655. However, other information about the territory was on the move at this time. Texts from the continent were finally being translated for English readers, whose penchant for discovery stories seemed to be growing stronger in this era.

John Ogilby's 1670 publication – *America: Being an Accurate Description of the New World* – was one of the first examples of this. It was a translation of *De Nieuwe en Onbekende Weereld*, by the Dutch author Arnoldus Montanus. Neither man had travelled to the locations described in the book but, instead, Montanus had based his work on an inconsistent collection of extant accounts, which he adapted with an eye fixed at times more on an interesting story than necessarily on troublesome considerations such as accuracy. Consequently, the work was strewn with errors and exaggerations. Ogilby, a translator and cartographer, was more fastidious when dealing with Montanus' text than Montanus had been when writing it. However, inevitably, the English edition of the book ended up being more of an impression in parts than an accurate rendition of the original texts because of Montanus' flexible treatment of them.

An extract in Ogilby's *America* relating to Tasman's encounter with a group of Māori gave most English readers their first glimpse of New Zealand's inhabitants. They could now draw a mental image of the people who occupied the place on the western extremity of their Moxon world map:

> The next day the *Southlanders* came in several Boats, two and two ty'd together, and cover'd with Planks, towards the Ships; the Gunner of the *Hemskerk* going with six Men in a Boat to help mount some Guns in the *Sea-Cock*, were betwixt both Ships set upon by the *Southlanders*, who approacht with a hideous noise, kill'd four of the *Hollanders* with long Poles, and forc'd the other three to save themselves by swimming; which they had no sooner done, but they Row'd with incredible swiftness towards the Shore, insomuch that they were out of the reach of their Guns before they could make ready to fire at them. *Tasman* finding that there was no good to be done here, since he hazarded the Lives of his Seamen against a company of wild People, he set Sail, but was follow'd by divers Boats, at which he fir'd his After-Guns with pieces of Iron and Stones, which killing some of them, made the rest return. In the midst of the Boats, which were ty'd two and two together, sat the Commander, who encourag'd the Rowers; they all us'd Clubs without Points.
>
> These People were gross of Body, undaunted, strong, and of a tawny colour; the Hair of their Heads stroak'd up round, was ty'd up on their Crown, on which stuck a stiff white Feather; about their Necks hung a square Plate; they wore in stead of Cloaks, square pieces of Cloth, made fast before on their Breasts.[1]

Although this book was generously illustrated (including several pictures of various mythical creatures) it contained no map or depictions of New Zealand. Instead, all that readers were offered was this crude caricature of Māori. If anything, though, this overdrawn vignette encouraged more curiosity among Britons in the world's most distant

1. J. Ogilby, *America: Being an Accurate Description of the New World; Containing the Original of the Inhabitants; the Remarkable Voyages Thither* (London: Thomas Johnson, 1670), p. 655.

extremities. The starkness in contrast between British civilisation and the savagery of other parts of the world was at once horrifying and deeply intriguing. Even the sparsity of detail, such as in the above account of Māori, encouraged British readers to plumb their imaginations to colour in the gaps.

Meanwhile, the map-printing business in England was growing. When visiting the Netherlands in 1663, the English compass manufacturer and map trader, John Seller, acquired some old copper plates of world maps,[2] which he began to recraft for the English market once back at his workshop in Wapping, near the Tower of London.[3] He published various maps in the following years and his career as a retailer of maps might have continued without attracting much attention, had he not been appointed 'Hydrographer to the Kings Most Excellent Majestie' in 1671. The following year, still primed with the prestige of the post, he produced the first English version of a map of the South Pacific. His *Chart of the South-Sea*[4] included an image of New Zealand in the bottom left-hand corner of the map. It contained more place-names than Moxon's map and terms such as 'bay', 'hills', 'point' and 'island' were all in English. As a sign that Seller had neglected to read widely when preparing his map, though, he still referred to the territory by its pre-1645 name, which he anglicised as 'States land'. However, one effect of his detailed maps of various parts of the world that he published throughout the 1670s was that it enabled English readers to perceive distant locations as being accessible, and foreign places and peoples as slightly more knowable.[5] Like Moxon's map (and all its predecessors), Seller's work portrayed New Zealand as an incomplete outline. In a way, this begged the question for those looking at the image: what was the shape of the missing area? It prompted the sort of speculative curiosity about the dimensions of the country that was unlikely to be resolved in the near future, however. As Seller's map informed viewers, the last European visit to New Zealand

2. J. Davis and C. Daniel, 'John Seller: Instrument Maker and Plagiarist', *Bulletin of the Scientific Instrument Society* 102 (2009), p. 6.
3. H. Wallis, 'The Eva G.R. Taylor Lecture: Navigators and Mathematical Practitioners in Samuel Pepys's Day', *Journal of Navigation* 47, no. 1 (January 1994), p. 7.
4. J. Seller, *A Chart of the South-Sea*, London (*c*. 1672), Boston Public Library, Norman B. Leventhal Map Center, Identifier: 06_01_008267, Call #:G1059 .S45 1672
5. P.E. Steinberg, 'Calculating Similitude and Difference: John Seller and the 'Placing' of English Subjects in a Global Community of Nations', *Social & Cultural Geography* 7, no. 5 (2006), p. 687.

was three decades ago, which was as much of a hint as anyone needed that expeditions to this part of the world were an extremely infrequent event.

So, while the mystery of New Zealand's outline looked set to remain for the foreseeable future, English readers' interest in the area had their curiosity partially satiated by the occasional appearance of texts which made some mention of the territory. Their next serving of information about New Zealand came in 1682, with the publication of Robert Hooke's article, *A Short Relation out of the Journal of Captain Abel Jansen Tasman*. Hooke was a polymath and had either dabbled or immersed himself variously in medicine, architecture mechanics, chemistry, physics, geology and mapmaking. In 1665 he was appointed Curator of Experiments at the Royal Society – a position that was bestowed on him for life. His eighteenth-century biographer, Richard Waller, summed up Hooke's temperament as 'melancholy, mistrustful and jealous',[6] yet his brilliance in so many fields made him in many ways a natural choice for his role at the society.

Amid the many activities that preoccupied him, Hooke surprisingly found time to write an abridged translation of the published version of Tasman's journal for the Royal Society in 1682.[7] However, instead of the work being released to the general public, it remained essentially an 'in-house' publication for members of the society. In the most precise language that he could summon, he relayed some of Tasman's encounters from their original Dutch:

> These Inhabitants were rough of voice, thick and gross made, they came not within a stones cast on board of us and blew several times on an instrument which made a noise like a moorish Trumpet, in answer thereto we blew ours. Their colour was between brown and yellow; they had black hair bound fast and tight upon the crown of their head, in the same manner as the *Japanners* have theirs behind their head, and near as long and thick of hair, upon which stood a great thick

6. R. Waller, *The Posthumous Works of Dr. Robert Hooke* (London: Sam Smith, 1705), p. xxvii.
7. E. Stokes, 'European Discovery of New Zealand before 1642: A Review of the Evidence', *New Zealand Journal of History* 4, no. 1 (1970), p. 18; S. Shapin, 'Who Was Robert Hooke?', in M. Hunter and S. Schaffer (eds), *Robert Hooke: New Studies* (Woodbridge, Suffolk: The Boydell Press, 1989), pp. 253-58.

> white Feather: their clothes were of Mats, others of Cotton but the upper parts were naked.
>
> ... these Antipodes began to be somewhat bolder, and more free, so that they indeavoured to ... Merchandize with the Yacht, and began to come on board; the Commander seeing this began to fear, lest they might be fallen upon, and sent his Boat or Prow with seven men to advertise them that they should not trust these people too much: they went off from the Ship, and not having any Arms with them, were set upon by these Inhabitants, and three or four of them were killed, & the rest saved themselves by swimming: this they indeavoured to revenge, but the water going high they, were hindred; this Bay was by them for this reason named *Murderers Bay*, as it is marked in the Charts.[8]

A few of the details, such as the white feathers worn on the heads of some Māori, matched precisely other translations of this account, but the narrative strayed off the path of previous versions at several points. There are also questions that hang over Hooke's motives for translating this brief extract of Tasman's journal. Why, for example, did he feel the need to produce this text when Moxon's version in English was already in circulation? Why did he only reproduce such a small quantity of Tasman's account of New Zealand? What was the purpose of publishing this primarily just for the readership of the Royal Society? Most puzzling of all, why focus on a segment of a journal written four decades earlier about a remote, small and apparently valueless territory? It could be that Hooke undertook the entire effort simply to demonstrate to himself that he had mastered another tributary of knowledge, however diminutive it may have been. It could also be that something about this partially known and incompletely mapped sliver of land lodged itself in some recess in his mind and became an intellectual itch that he wanted to scratch. Maybe, through drawing the attention of other members of the Royal Society to this far-off location, he hoped that their interest in the territory would similarly start to tingle. There was one factor, however,

---

8. R. Hooke, 'A Short Relation out of the Journal of Captain Abel Jansen Tasman, upon the Discovery of the South Terra Incognita: not long since Published in the Low Dutch by Dirk Rembrantse', *Philosophical Collections* (London: R. Chiswell, 1682), p. 181; G.R. Crone, 'The Discovery of Tasmania and New Zealand', *Geographical Journal* 111, nos 4/6 (1948), pp. 258-61.

that deserves some consideration when assessing why Hooke might have gone to the trouble of producing this extract: the brooding presence of the English government in the workings of the Royal Society.

In 1660, when the society was formed, some onlookers may have regarded it – quite unfairly – as little more than an association of educated hobbyists. However, within a decade, the virtuosic credentials of many of its members, along with the advances in various sciences that they were achieving, gave cause for officials and politicians to reconsider any earlier cynicism they may have had and to regard the society's activities much more seriously. By the 1680s, the influence of the English government was slowly beginning to be felt in the society's sinews, even if it was not especially evident to outsiders.

Initially, support for the society came not directly through government policy or statutes, but through the softer and less conspicuous approach of patronage and influence. This was achieved partly though the composition of the society's membership, of which roughly a quarter by 1685 were connected with the state, either as politicians, diplomats, courtiers or officials.[9] The effects of such overlapping relationships seldom leave firm traces on documents, though, and so it can be difficult to establish with any certainty the strength of government influence over the society at this time.[10] It would, however, be naïve to believe that the society was above any official sway at all.

There were occasional exceptions, though, to this obscured relationship, such as the government's overt support of the Royal Society's programme for mapping England in the 1680s,[11] its encouragement of efforts by some society members to establish the 'wealth, growth, trade, and strength of the nation'[12] and a request by the Admiralty for the society to provide its expertise for raising some naval ships that had sunk at Woolwich.[13] What these and other examples illustrate is

9. J. Gascoigne, 'The Royal Society and the Emergence of Science as an Instrument of State Policy', *British Journal for the History of Science* 32, no. 2 (1999), p. 171.
10. M. Hunter, 'The Crown, the Public and the New Science, 1689-1702', *Notes and Records of the Royal Society of London* 43 (1989), 99-116.
11. P. Slack, 'Government and Information in Seventeenth-Century England', *Past & Present* 184, no. 1 (2004), p. 37.
12. C.H. Hull (ed.), *Economic Writings of Sir William Petty*, 2 vols (Cambridge: Cambridge University Press, 1899), Vol. 1, p. 95, in Slack, 'Government and Information in Seventeenth-Century England', p. 37.
13. C.R. Weld, *A History of the Royal Society, with Memoirs of the Presidents*, 2 vols (London: John W. Parker, 1848), Vol. 1, p. 2321.

that the liaison between the society and the government was growing throughout this period – both formally and informally[14] – and that it was far more serious by the end of the century than it had been at its flirtatious stage in the 1660s.

As cosy and as mutually rewarding as this relationship was becoming, it was by no means exclusive. The English East India Company, which was still in the foothills of its own imperial ascent at this time, had obtained a renewed Royal Charter in 1661 (a year before the Royal Society)[15] and was busily expanding its trade networks internationally. Sometime around 1687, Thomas Bowrey – a trader associated with the EIC – produced a map which contained the now-familiar outline image of New Zealand (although even at this relatively late date, he labelled it 'States Land').[16] Bowrey's various commercial enterprises almost all ended in ruin,[17] but he had a persistent interest in the places he travelled to, as well as those that he had found out about from other sources. His map on which New Zealand appeared was neither named nor dated, but the type of paper it was drawn on, the ink that was used and the handwriting, all point to Bowrey being its creator.[18] Of course, 'creator' in this context does not equate with originator, as Bowrey's map is plainly a copy,[19] based probably on a reproduction of Visscher's map, or another early version which he was shown during the period when he was trading in South East Asia (possibly in Madras).[20]

---

14. Burney, *A Chronological History of the Discoveries in the South Sea*, Vol. 3, p. 316.
15. Charters of the East India Company with related documents: the 'parchment records', British Library, Asian and African Studies, ref. IOR/A/1.
16. T. Bowrey, Correspondence, diaries, drawings, charts, maps and other papers, London Metropolitan Archives, CLC/427/MS24176 (microfilm MS24177/001).
17. S. Paul, *Jeopardy of Every Wind: The Biography of Captain Thomas Bowrey* (London: Monsoon Books, 2020), pp. 169-219.
18. These matching details are in a Bowrey map from at Fort St George in Chennai, India. R. Henry (ed.), *Early Voyages to Terra Australis, Now Called Australia: A Collection of Documents, and Extracts from Early Manuscript Maps* (London: Hakluyt Society, 1859), pp. xcvi-xcvii.
19. Ibid.
20. G.R. Crone, 'The Discovery of Tasmania and New Zealand', *Geographical Journal* 111, nos. 4/6 (1948), pp. 258-61; B. Hooker, 'New Light on Jodocus Hondius' Great World Mercator Map of 1598', *Geographical Journal* 159, no. 1 (1993), pp. 45-50.

The significance of Bowrey's map lay not so much in its origin or form as its destination. The English East India Company had previously received intelligence of Tasman's expedition shortly after its completion, courtesy of one of its officials in Bantam. When that note had reached London, though, it was glanced at quickly, deemed to be of no value and filed away. The company was a very different entity at that time. Its finances were fickle and it was backed by slightly shady speculators – men who were prepared to gamble their capital for the prospects of disproportionate financial returns. In a way, the company gave the impression of a get-rich-quick scheme and, when the profits did not reach investors as swiftly as they had hoped, most had no option but to wait.

In the decades that followed, the company grew, with trading posts peppered throughout parts of Southeast Asia. Its structure became more formal, its bureaucracy more efficient and its dividends increased, both in amount and frequency of payment. Most of all, however, from the end of the seventeenth century, the company became more territorially ambitious.[21] Thus, by the time that Bowrey's map portraying New Zealand reached EIC headquarters, it fell into that category of intelligence that was now treated much more seriously.[22] The other vital distinction that added value to Bowrey's chart was that, unlike the maps of New Zealand that appeared and were reproduced within the realm of the Royal Society in this period, it depicted the specific route that Tasman had used when sailing from the Indian Ocean to New Zealand in late 1642. If it was not to be used straight away, at least such information was now kept at hand in the event that future commercial imperatives might cause the company to revisit the map and the location that it depicted.

Meanwhile, the Royal Society, looking to shore up its own importance and relevance, in 1694, published extracts of records from several European expeditions to various parts of the world in its *An Account of Several Late Voyages and Discoveries*. Tasman's journal covering his expedition to Australia and New Zealand was included in this collection (although without any accompanying maps – these were promised to

21. D. Massarella, 'Chinese, Tartars and "Thea" or a Tale of Two Companies: The English East India Company and Taiwan in the Late Seventeenth Century', *Journal of the Royal Asiatic Society* 3, no. 3 (1993), pp. 399-403.
22. A. Winterbottom, 'Producing and Using the Historical Relation of Ceylon: Robert Knox, the East India Company and the Royal Society', *British Journal for the History of Science* 42, no. 4 (2009), p. 533.

be printed soon).[23] The segment of Tasman's account appearing in this work offered readers nothing new: it had been lifted word for word from Hooke's translation that had been published twelve years earlier. Yet, the Royal Society disingenuously claimed that its 1694 publication revealed details about Tasman's journey that were 'not yet known to the English'[24] – a statement that was demonstrably false and which anyone with more than a passing interest in the literature on New Zealand in this era would have at once realised.

Why, then, would the Royal Society risk such reputational damage? Part of the reason is that works such as *An Account of Several Late Voyages and Discoveries* arguably were not produced simply as a source of information. Instead, they were intended to promote the image of the society as a progenitor and propagator of knowledge. If you were looking in England at the close of the seventeenth century for an authority on all things to do with the world's geography, the Royal Society was jostling to ensure that it would be the institution that embodied that authority. Moreover, as if to emphasise the point, this particular volume (which just happened to be issued by the society's own printer) was dedicated to Pepys (who had served as the society's president from 1684 to 1686), in acknowledgement of his 'Kindness and Generosity to the Publick ... in advancing the Progress of Useful Knowledge'.[25] The Royal Society was fast emerging as the (self-appointed) font of all scientific knowledge in the nation.

Nonetheless, when it came to any new cartographic information about New Zealand, there simply was none and had not been any for half a century. It would be mistaken, though, to assume that no additional intelligence on the territory somehow equated with interest in the region falling dormant. On the contrary, the absence of fresh details about New Zealand over a period of several decades led to a mounting curiosity about the territory, and especially over its missing dimensions – but, as there were no means of satisfying this inquisitiveness for the moment, it remained in suspension, waiting for a resolution.

Awareness of New Zealand's location was reaching an ever-broader public audience. It was not that this led to the clamour of crowds, impatient to know more and demanding that an expedition be despatched to the

---

23. The Royal Society, *An Alphabetical Catalogue Abreviated in the Philosophical Transactions* (London: Sam Smith, 1694), pp. 167-68.
24. S. Smith, *An Account of Several Late Voyages and Discoveries to the South and North* (London: Sam Smith, 1694), p. xxvii; Burney, *A Chronological History of the Discoveries in the South Seas*, Vol. 3, pp. 316-17;
25. Smith, *An Account of Several Late Voyages*, p. i.

region forthwith. Rather – as the owners of maps or atlases leisurely surveyed the details of the world's layout in a low chair in their reading rooms, perhaps while sipping a cup of that fashionable new drink, tea (which the English East India Company commenced exporting from Bantam in 1667)[26] – the idea of completing the unfinished business of mapping the outlines of all the world's territories nestled among their mental furniture and remained there. Even if no expedition to these remote regions was imminent, there would still be new information (or more likely, rehashed existing information) about them trickling through to readers in England; and, with each drop of intelligence that entered the nation's general knowledge, the weight of that fascination and expectation grew.

'Creative' Mapmaking

By the early eighteenth century, the pool of new information about New Zealand had effectively dried up. Yet, British readers were being inundated with a stream of maps containing images of the territory. In 1700, the one-time sea captain, and now mapmaker and publisher, John Thornton, released his *Atlas Maritimus* – which included what he advertised as 'A New Map of the World'.[27] Thornton had received some cartographical training from John Seller and, on the latter's death in 1697, had taken over a portion of his business. The more important aspects of his career involved his work as hydrographer both for the Hudson Bay Company and the English East India Company,[28] during which time he assembled the maps for *Atlas Maritimus*. New Zealand was not mentioned in the text of the work but it appeared, as Zelandia Nova, on the world map that he included in the book.

The place-names that Thornton applied to the territory were selected from previous Dutch maps but included some, rather than all, of the names that existed in earlier charts. What sets Thornton's rendition of New Zealand apart from its predecessors was that the contours of the coast were much more dimpled, with bays, inlets and peninsulas, giving

26. J. Macgregor, *Commercial Statistics: A Digest of the Productive Resources, Commercial Legislation, Customs Tariffs, of All Nations. Including All British Commercial Treaties with Foreign States* (London: Whittaker & Co., 1850), Vol. 3, p. 47.
27. J. Thornton, *Atlas Maritimus or, The Sea-Atlas: Being a Book of Maritime Charts* (London: n.p., 1700), p. 11.
28. M. de La Roncière, 'Manuscript Charts by John Thornton, Hydrographer of the East India Company (1669-1701)', *Imago Mundi* 19 (1965), p. 46.

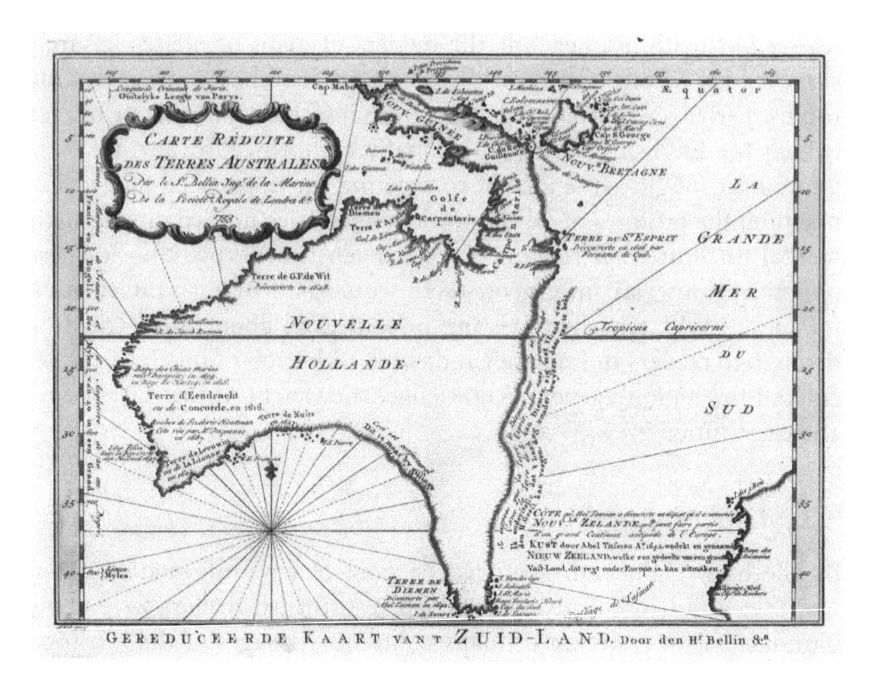

Carte réduite des terres Australes (Jacques Nicolas Bellin, 1753)
Stanford Libraries, Barry Lawrence Ruderman Map Collection

the impression of greater precision, even though this was an exercise in pure imagination. Indeed, the result was to render this stretch of the territory even more inaccurate than it had been depicted on any other previous map.

Two years later, William Godson produced a hand-coloured wall map (engraved and sold by the London mapmaker George Willdey) which was advertised as:

> A new and correct map of the world: laid down according to the newest observations & discoveries in several different projections including the trade winds, monsoons, variation of the compass, and illustrated with a coelestial planisphere, the various systems of Ptolomy, Copernicus, and Tycho Brahe together with ye apearances of the planets &c.[29]

29. W. Godson, *A New and Correct Map of the World: Laid down according to the newest observations & discoveries in several different projections including the trade winds, monsoons, variation of the compass, and illustrated with*

However, despite such promising credentials, information on New Zealand (spelt 'New Zeland' on this map) was more jumbled than on most other contemporary portrayals of the territory. Cape Maria van Diemen was anglicised and contracted to Cape Mary; what would later become the North Island was misnamed Kings Island (which obviously was mistakenly derived from the Three Kings Islands that in Tasman's map clearly lay to the north of this body of land); Murderers' Bay was now Assassins' Bay; and there were curious references to Tasmans Road and Sand Hill – neither of which appeared in that form on earlier maps. However, as with most other world maps produced at this time, the constraints of space, coupled with most attention being lavished on Europe, the Americas and Southeast Asia, meant that New Zealand ended up being engraved as a sort of geographical accessory. It was included because its location in that space was known, but it was hardly a commercial imperative to have the territory's depiction representing the epitome of exactitude.

As Thornton had discovered, the near-straight lines that had appeared on some maps in the previous century representing New Zealand could be embellished to at least give the impression of detail and, therefore, accuracy. Such an approach was also taken up by John Harris in 1705, in his sprawling book, *Navigantium atque Itinerantium Bibliotheca*. The map of New Zealand which it contained (with the territory referred to as Zeelandia Nova) resembled Thornton's map, with the outline of the coast appearing even more finely detailed.[30] Any reader examining this image would understandably presume that it was based on a close survey of the coast by experienced cartographers, rather than the speculative lineal ramblings of a mapmaker, but, as long as there was no way of anyone knowing any better, mapmakers like Thornton and Harris could get away with their more creative approach to representing territories such as New Zealand.

A revised version of Harris' book appeared in 1744 – a full 25 years after his death. This updated edition was probably made by John Campbell and now contained an abridged account of Tasman's journal, along with a map of New Zealand. According to an accompanying note, this map had been produced by the Welsh engraver Emanuel Bowen, who

---

a coelestial planisphere, the various systems of Ptolomy, Copernicus, and Tycho Brahe together with ye apearances of the planets &c. (London: George Willdey, 1702).

30. J. Harris, *Navigantium atque Itinerantium Bibliotheca* (London: Thomas Bennet, 1705), p. 324.

had copied it more or less directly from one in the possession of the Dutch East India Company in Amsterdam.[31] However, the exaggerated coastal outline of New Zealand that appears on the map suggests it is more probable that this was copied from recent English maps of the region. The reference to Dutch maps was most likely an appeal to the authority of Amsterdam's cartographers, whose reputation throughout Europe had reached its zenith in this era, and it was the records of Dutch merchants and explorers which occupied a good portion of *Navigantium atque Itinerantium Bibliotheca.*

However, while acclaim for Europe's accumulation of geographical knowledge in the previous century no doubt belonged primarily to the Netherlands, by the 1700s, this status was facing a challenge. The British no longer wished to remain a technological, economic, military, let alone intellectual, poor cousin to their continental competitors, and an intimation of this sea change in attitude appeared in the fine-print of Bowen's map. The passage of text – inserted in a legend positioned in the expanse of the Tasman Sea – referred to Australasia and proposed that 'whoever perfectly discovers & settles it will become infalliably [*sic*] possessed of Territories as Rich, as fruitful, & as capable of Improvement, as any that have been hither to found out, either in the East Indies, or the West'.[32] In this case, that 'whoever' plainly hinted at Britain itself – a nation that had officially come into being by the Act of Union 1707.

This was a crude recommendation in some respects: claim the territory; extract the wealth from it; and everyone will be better off (apart from the indigenous inhabitants, of course, but then, who worried about them?). In his enthusiasm, Bowen also overlooked the fact that another great European maritime power had been responsible for mapping and exploring these very territories in the previous century and had remained steadfastly uninterested in them ever since. This alone ought to have undermined Bowen's imperial ardour and made him think twice before advocating Britain's colonisation of Australia and New Zealand.

However, Bowen was far from alone in harbouring global ambitions for his nation. Since the beginning of the eighteenth century, British

31. E. Bowen, *A Complete Map of the Southern Continent Survey'd by Capt. Abel Tasman & depicted by Order of the East India Company in Holland in the Stadt House at Amsterdam* (London, 1744), Stanford Libraries, Barry Lawrence Ruderman Map Collection.
32. J. Harris, *Navigantium atque Itinerantium Bibliotheca , or, A Complete Collection of Voyages and Travels*, ed. John Campbell (London: T. Woodward, 1744), p. 324.

intellectuals, politicians, clerics and merchants had been lining up to make their case that the arc of history was now bending in their country's favour and that imperial expansion was not just an intriguing option, but practically a national entitlement. The groundwork for such expectations had been laid in the latter seventeenth century, but now it was time to build the edifice of empire upon it. The poet James Thomson captured some of this sentiment when he unblushingly claimed in 1740 that Britain's imperial destiny was divinely ordained and would eventually radiate in all directions:

> *As navies grow, as commerce swells her sail*
> *With every breeze that under heaven can blow*
> *From either pole; thro worlds yet unexplor'd,*
> *In east and west, that to thy sons disclose*
> *Their golden stores, their wealth of various name,*
> *And lavish pour it on Britannia's lap!*[33]

This was an empire 'whose bounds nature has not yet ascertained',[34] one official concluded, and which a cleric proposed in 1759 was a 'Birthright' of Britons, obliging them 'to carry not only Good Manners, but the purest Light of the Gospel, where Barbarism and Ignorance totally prevailed'.[35] The result, according John Campbell, was 'a very noble and shining Instance of that Prosperity', which was derived from the nation's 'Possessions, Colonies, and Settlements in all different Parts of the Globe', that had 'contributed to the Grandeur and Opulence of the British Empire'. It was not just Britain, though, that was reaping imperial dividends. Campbell claimed that its 'Dominions so extensive and at so great a Distance have been acquired and united to us by the Ties of mutual Interests and a reciprocal Communication of Benefits'.[36]

---

33. J. Thomson, *Alfred: A Masque* (London: A. Millar, 1751), p. 63.
34. G. Macartney, *An Account of Ireland in 1773: By a Late Chief Secretary of that Kingdom* (London, n.p., 1773), p. 55.
35. R. Brewster, cited in K. Wilson, 'Empire of Virtue: The Imperial Project and Hanoverian Culture *c.* 1720-1785', in L. Stone (ed.), *An Imperial State at War: Britain from 1689 to 1815* (London: Routledge, 1994), p. 128, in J.P. Greene, 'Empire and Identity from the Glorious Revolution to the American Revolution', in P.J. Marshall (ed.), *The Oxford History of the British Empire: The Eighteenth Century* (Oxford: Oxford University Press, 1998), p. 219
36. J. Campbell, *A Political Survey of Britain: Being a Series of Reflections on the Situation, Lands, Inhabitants, Revenues, Colonies, and Commerce of this Island* (London: Richardson & Urquhart, 1774), Vol. 1, p. iv.

This was envisaged not as a regime of tyranny, but one of reciprocal rights and rewards. These cheerleaders for British expansion during this period also believed that, in addition to being more lucrative and on a grander scale, theirs would also be a 'better' empire than any of its predecessors. It would be Protestant, integrated by commerce rather than force, would be connected by oceanic rather than territorial routes and, above all, the ascendancy of British law, liberties and property rights would make it an empire of liberty.[37]

The one monopoly that proved perpetually elusive for the British, however, was intelligence. Mapmaking in London was still a cottage industry (literally, in a few cases) in this era, whereas in the European capitals of cartography, especially Amsterdam and Paris, it was carried out on a larger and altogether more professional scale. The Dutch, in particular, were at an advantage because they still held the lion's share of raw material relating to explorations in various parts of the world over the preceding 150 years and because the scale of their cartographical activities had reached industrial dimensions.

One of the most exhaustive uses of Dutch East India Company records was a multi-volume work by François Valentyn (or Valentijn), published between 1724 and 1726. This collection, which he entitled *Oud en Nieuw Oost-Indiën*[38] ('Old and New East Indies'), extended to around 5,000 pages and was intended to be a detailed (which it certainly was) and accurate (which it mostly was) reference work, drawing togther as much relevant information about the Dutch Empire as was available.[39] Inevitably, some errors crept their way into this book (including confirmation that mermaids existed in the southern hemisphere – something Valentyn claimed to have witnessed personally in May 1714)[40] but overall it was the most thorough single summation of Europe's knowledge of the non-European world in existence in the early

37. D. Armitage, *The Ideological Origins of the British Empire* (Cambridge: Cambridge University Press, 2000), p. 8.
38. F. Valentijn, *Oud en Nieuw Oost-Indiën, vervattende een naaukeurige en uitvoerige verhandelinge van Nederlands mogentheyd in die gewesten, benevens eene wydluftige beschryvinge der Moluccos, Amboina, Banda, Timor, en Solor, Java, en alle de eylanden onder dezelve landbestieringen behoorende, het Nederlands comptoir op Suratte, en de levens der Groote Mogols* (Dordrecht: By Joannes van Braam, Gerard Onder de Linden, 1724-26).
39. S. Huigen, 'Repackaging East Indies Natural History in François Valentyn's *Oud en Nieuw Oost-Indiën*', *Early Modern Low Countries* 3, no. 2 (2019), 234-264.
40. F. Valentijn, *Oud en Nieuw Oost-Indiën*, p. 331.

eighteenth century. Valentyn had an advantage over his contemporaries in that he had lived in the VOC's settlement in Java for a total of sixteen years, where he was employed as a chaplain. During this period, he amassed a vast collection of documents which he began to work on for his book once he returned to the Netherlands.

Amid the thousands of pages of text and hundreds of illustrations in Valentyn's volumes were two engravings of the coast of New Zealand (a territory referred to simultaneously as N. Zeeland and Staaten Land on one of the maps). Unusually for a publication by one author, there is significant variation between these two maps. The first one (part of a world map) contains a highly distorted rendition of Tasman's chart of the South Island.[41] The second map, in contrast, is much more in keeping with the details contained in Tasman's journal and with the sketch of the country's coast produced during his voyage there in 1642. Adjoining the second map is a picture of Tasman's visit to the Three Kings Islands, depicting two Māori gesticulating towards his ships.[42] This image is obviously based on the texts of the expedition's journals as no such scene was produced by any of the crew at the time.

What this second map strongly hints at is that Valentyn had access to a copy of Tasman's map during his time in Java (Valentyn first arrived on the island 40 years after Tasman had returned from his 1642-43 expedition). This makes it probable that a direct copy of Tasman's map (or even an original version) was carefully stored in the VOC archives in Batavia, and that it was accessible to people who wished to use it for scholarly rather than just commercial purposes. This level of access, particularly with the knowledge that Valentyn intended to publish his research, is also good indication that Dutch authorities in Southeast Asia had completely abandoned any attempts to keep such intelligence confidential, and that this position was endorsed by the company's directors in Amsterdam. It is therefore surprising that British cartographers continued knowingly to produce copies of copies of maps of New Zealand, when from 1726, they had access to Valentyn's detailed map of the territory, which was most likely obtained directly from the Dutch East India Company archives in Batavia. The most likely explanation for this is also the least appealing one: that New Zealand was too inconsequential commercially, too diminutive territorially and too remote geographically to matter when it came to the precision with which it was portrayed on maps – an attitude that even extended to the

41. Ibid., p. 46.
42. Ibid., p. 52.

names that Tasman had given to some of the locations he had charted along its coast which were omitted from several subsequent maps printed elsewhere in Europe. Furthermore, as each decade passed without any European power revisiting New Zealand following Tasman's expedition, the impression of the territory's insignificance was only reinforced.

It turned out, though, that the relentless upward march of Enlightenment knowledge was not as sure-footed as many historians later liked to see it. There were occasional missteps, of which the alleged works of François [Francisco] Coreal was a glaring example. Alleged because some of the material he wrote was 'fake'[43] and because he most likely never existed. Instead, the work was probably apocryphal and the name was a pseudonym 'for somebody who never undertook the voyages described'.[44] The book in question was *Voyages de François Coreal aux Indes Occidentales*, published in Paris in 1722.[45] It contained a small fold-out map of the world with New Zealand appearing on it – not that a reader would know this from looking at the map. There was no country name attached to the territory, only three local names: Cape Maria van Diemen; Three Kings Islands; and Tasman Harbour (actually Cook Strait). For all its shortcomings, though, this work served as the most popular source from which subsequent writers throughout Europe obtained information about Tasman's expedition.[46] Such was the state of map reproduction and data dissemination in this era. There were so many incarnations of images of New Zealand from which to draw that it is impossible to know which particular map or maps were examined by cartographers before they created their own impression of the territory. Distortions by cartographers were inevitable and, for readers, unknowable. If you purchased one of these expensive volumes in the mid-1700s, you had little reason to doubt its veracity – especially when even

43. P.G. Adams, *Travel Literature and the Evolution of the Novel* (Lexington: University Press of Kentucky, 1983), p. 210.
44. Sotheby's, *Voyages de François Coreal aux Indes Occidentales ... nouvelle edition* (Paris: Robert-Marc d'Espilly, 1722), L12401, Lot 121, catalogue note. Also see Library of Congress, 'Voyages de François Coreal aux Indes Occidentales ...', notes, Library of Congress Control Number 02000067.
45. F. Coreal, *Voyages de François Coreal aux Indes Occidentales, contenant ce qu'il y a vû de plus remarquable pendant son séjour depuis 1666 jusqu'en 1697: traduits de l'espagnol. Avec une relation de la Guiane de Walter Raleigh, & le voyage de Narbrough à la mer de Sud par le détroit de Magellan, &c* (Paris: Chez André Cailleau, 1722).
46. McCormick, *Tasman and New Zealand*, p. 19.

the professionals producing these books and maps were not moored to any fixed and universally agreed-on acceptance of New Zealand's shape. To some extent, the trend away from earlier cartographical styles – with their ships, sea monsters, Classical motifs and florid decorations – towards a plainer, more clinical appearance, was misleading. Yes, they looked more scientific, but this did not necessarily mean they were any more accurate.

One of the great European cartographers of this period was Jacques Nicolas Bellin, who in 1741 was appointed Official Hydrographer of the French king.[47] Although firmly ensconced in his work, and esteemed by his Parisian colleagues, Bellin looked to Britain, and specifically its Royal Society, when he decided he wanted to be affiliated with an academic institution which he felt befitted his abilities and standing. It was an unusual and potentially uncomfortable decision for him to make, because being a government cartographer meant that he was unavoidably part of the machinery of the French navy. Thus, any association with Britain could possibly have some security implications, especially given the relations between the two nations, which so often chafed on contact. However, if there were any concerns, they were apparently insufficient to prevent Bellin becoming a member of what was probably the continent's leading scientific organisation from the second half of the eighteenth century.[48]

Through his association with the Royal Society, Bellin began acquiring copies of several recently-made British maps, sea charts and even the journals of some British navigators, which no doubt informed his own mapmaking work. In 1748, a world map he produced appeared in *Histoire générale des voyages*[49] – a meandering fifteen-volume work by Antoine François Prévost (Abbé Prévost), a French novelist, sometime monk and author of *Manon Lescaut* (which formed the basis of the libretto for Puccini's 1892 opera of the same name). Bellin's map showed that even the most proficient cartographers in France could succumb to the errancy that afflicted so many previous renditions of New Zealand.

47. Zandvliet, 'Mapping the Dutch World Overseas', p. 1444.

48. M. Pastoureau, 'Jacques-Nicolas Bellin, French Hydrographer, and the Royal Society in the Eighteenth Century', *Yale University Library Gazette* 68, nos 1/2 (1993), pp. 66-67.

49. A. Prévost, *Histoire générale des Voyages, ou, Nouvelle collection de toutes les relations de voyages par mer et par terre ... composées sur les observations les plus autentiques* (Paris: Didot, 1746-59), Vol. 6, insert.

It is probable that Bellin referred to maps he had acquired during his trips to London and so the practice of making and modifying a copy of a copy of a copy went for yet another round.

What makes Bellin's map mystifying, however, is that five years after it appeared in Prévost's work, he produced another version. Normally, as had been the practice with almost all other cartographers for more than a century, revised editions of maps involved making minor modifications to the existing copper plates from which they were printed – and, in the absence of any new information, most of the content of the maps would remain largely untouched.

However, for some reason, Bellin radically altered the shape of New Zealand on his revised map.[50] Not only was the coast now almost completely unrecognisable from that which Tasman had drafted, or indeed which any subsequent cartographer had drawn, but, for the first time, the territory was no longer depicted simply as a single line of coast. Instead, he enlarged the scale of New Zealand drastically. From the bottom section of Tasman's original portion of the territory's coast, Bellin extended a speculative coast extending south-east for at least 300 kilometres (before disappearing into the map's margin). From Cape Maria van Diemen, at the top of the country, the coast stretched hypothetically more than 500 kilometres north-east, before reaching the map's right-hand border. The only indication that these enlargements were not based on any prior map, or other source of evidence, was that the projected additional lengths of coast appeared in a marginally lighter shading. Bellin's justification for this colossal imaginative leap in redefining New Zealand's shape and scale represented another staggering misjudgement. On the map's legend, he made the claim that this territory was the Great Southern Continent – something that cartographers since Tasman had almost unanimously refuted. Bellin may have been trying to impress his colleagues in the Royal Society by reviving the theory of the Great Southern Continent and giving part of it a suggested form. However, whatever the motive, his 1753 map of New Zealand remains the least accurate of any produced of the territory and represents an ill-fated digression in the history of mapping that part of the world.

Perhaps other cartographers knew that Bellin had overstepped the mark on this occasion. When one of his compatriots, Robert de Vaugondy, published a map three years later, which also contained an image of New

---

50. J.N. Bellin, *Carte réduite des terres Australes* (Paris, 1753), Stanford Libraries, Barry Lawrence Ruderman Map Collection.

Zealand,[51] the form of the territory had reverted to the conventionally confused one that had appeared in some of the more recent attempts to depict the place, with its coast overwrought with detail and the usual list of place-names (this time, translated into French) appended to it. Admittedly, it was an inaccurate likeness of Tasman's map, but at least it was consistent with the inaccuracies in preceding images of New Zealand.

Newspapers and Travel Literature Feed Public Consciousness

Edmund Burke sensed it. From the middle of the eighteenth century, Britain (presumably with the rest of Europe trailing behind it) was crawling out from under the shadow of superstition and stale inherited certainties and into the intellectually invigorating glow of the Enlightenment. He wrote exuberantly in 1756:

> The Fabrick of Superstition has in this our Age and Nation received much ruder Shocks than it had ever felt before; and through the Chinks and Breaches of our Prison, we see such Glimmerings of Light, and feel such refreshing Airs of Liberty, as daily raise our Ardor for more. ... We begin to think and to act from Reason and from Nature alone.[52]

The acquisition of knowledge for its own sake had gone from being an exclusive indulgence to a widespread virtue.

Seemingly everywhere, people were busy collecting, classifying, curating and categorising all types of information about the world. Knowledge was expanding into every crevice which curiosity was opening up, instead of being cloistered in places like universities, where in previous generations it had become a rarefied commodity monopolised

51. R. de Vaugondy, *Carte réduite de l'Australasie, pour servir à la lecture de l'histoire des terres Australes [cartographic material] / par le Sr. Robert de Vaugondy, Geog. ord. du Roi, de l'Academie Royale des Sciences et Belles-Lettres, de Nancy, 1756; [engraved by] G. de-la-Haye*, National Library of Australia, Bib ID 1190483. Also see C. de Brosses, *Histoire des navigations aux terres australes* (Paris: Chez Durand, 1756), insert.
52. E. Burke, *A Vindication of Natural Society: or, A View of the Miseries and Evils Arising to Mankind* (London: M. Cooper, 1756), p. 8.

by academics. Gentlemen (and very occasionally, lady) scholars, along with all manner of amateur enthusiasts across the country, were forming an array of learned societies in this period – organisations where they could share insights and delve deeper into understanding the material world. These were people who were guided both by curiosity and by a sense of obligation to subsequent generations. 'Every rational Being should, nay, is obliged to bequeath something to Posterity', one disciple of the Enlightenment melodramatically wrote at the time, 'that it may be known there was once such a person who intended to prevent the destruction of *Human Knowledge*, from the Sithe of Time; and to *Eternize* the Memory, or *Actions* of all such Men as have signalized themselves in Merit.'[53] When plans were being made for a new library to be built in London, the classical scholar Richard Bentley hoped that, once completed, 'Societies may be formed, that shall meet, and have Conferences there about matters of Learning ... what Advantage and Glory may accrue to the Nation, by such Assemblies not confined to one Subject, but free to all parts of good Learning'.[54]

This blossoming interest in forms of organised, collective and democratic learning about specialist topics was made possible in part because of the rapid rise in literacy from the eighteenth century onwards (especially among urban middle-class males living in towns and cities). Newspapers, pamphlets, magazines and books were being churned out in increasing quantities and circulated with growing momentum[55] to feed the voracious appetite for information. Travel books, in particular, captivated the new reading classes,[56] whose increased spending power[57] enabled them to acquire such works on a more regular basis.

53. J. van Rymsdyk and A. van Rymsdyk, *Museum Britannicum: Being an Exhibition of a Great Variety of Antiquities and Natural Curiosities, Belonging to that Noble and Magnificent Cabinet, the British Museum* (London: I. Moore, 1778), p. i.
54. R. Bentley, in T.A. Bartholomew, *Richard Bentley, DD: A Bibliography of His Works and of All the Literature Called Forth by His Acts or His Writings* (Cambridge: Bowes & Bowes, 1908), p. 95.
55. H.T. Dickinson, *The Politics of the People in Eighteenth-Century Britain* (Basingstoke: Macmillan, 1994), p. 7.
56. M. Hunt, 'Racism, Imperialism, and the Traveler's Gaze in Eighteenth-Century England', *Journal of British Studies* 32, no. 4 (1993), p. 335.
57. E. Buringh and J.L. van Zanden, 'Charting the "Rise of the West": Manuscripts and Printed Books in Europe, A Long-Term Perspective from the Sixth through Eighteenth Centuries', *The Journal of Economic History* 69, no. 2 (2009), 409-45.

A sense of how fascinated Britons were with this genre of writing was captured by James Boswell, the dutiful biographer of Samuel Johnson, who recorded on one occasion towards the end of the century that travel literature was 'the common topick of conversation in London at this time, wherever I happened to be';[58] and, when it came to geographical information, by far the most accessible form in Europe in this era was the map[59] (which goes some way to explaining why cartographers were now enjoying such a flourishing trade in places like London). The actual numbers of maps printed in Britain in the eighteenth century remains elusive but, occasionally, a small indicative detail has survived the centuries that hints at the scale of production. As one example, in 1740, *The Gentleman's Magazine* noted that 20,000 copies of just one version of a map of the West Indies had been sold[60] (at which point, the copper plate from which it was printed would have begun to wear down to the extent that it would have noticeably affected the quality of the map). This must only have been a fraction of the maps produced in Britain. When, for example, Pepys died in 1703, this civil servant who was comfortable but not wealthy left behind a collection of around 1,100 maps.[61] Pepys was someone whose interest in cartography was not especially pronounced and, yet, like potentially thousands of others in Britain in a similar financial position, acquiring maps was almost part of gentlemanly etiquette.

It is evident that maps were a significant proportion of the printed works purchased in Britain in this century but their role went beyond a merely informative function. They had become part of the ensemble of knowledge that the educated could put on display to parade their learning and were even candidly marketed as an 'ornament' for gentlemen to exhibit in their houses.[62] Wall maps or globes were an obligatory furnishing in most schools from this time and were beginning to feature

58. J. Boswell, in M. Morris (ed.), *Boswell's Life of Johnson* (London: Macmillan & Co., 1899), p. 291.
59. M.S. Pedley, *The Commerce of Cartography: Making and Marketing Maps in Eighteenth-Century France and England* (Chicago: University of Chicago Press, 2005), p. 1.
60. Williamson, *British Masculinity in* The Gentleman's Magazine, p. 48; *The Gentleman's Magazine* 10 (London, January 1740), p. 3.
61. P. Koyoumjian, 'Ownership and Use of Maps in England, 1660-1760', *Imago Mundi* 73, no. 1 (2021), p. 36.
62. T. Porter, *A New Booke of Mapps* (London: Robert Walton, 1655), pp. 71-72; Koyoumjian, 'Ownership and Use of Maps in England, 1660-1760', p. 38.

in a variety of distinctly non-geographical places: ladies' fans, fireplace screens; curtains; handkerchiefs; and even watch cases.[63]

* * *

When considering the accelerating production-line of publications in Britain in the eighteenth century dealing with aspects of geography, it can be easy to overlook the self-evident point that these works were practically all in English. Place-names were anglicised, measurements were made using units that were familiar to Britons, navigational coordinates were included based on British landmarks, British trading posts overseas were labelled and some territories were classified according to their principal exports to Britain.[64] Inevitably the combination of these developments gave rise to a sense of geographic dominion over certain parts of the world in the minds of some Britons. It was as if a visual impression of the world was a precursor to more meddlesome forms of interest.

It was not just the colonisers who would benefit from any forthcoming involvement in other territories, though. In the 1750s, the Scottish philosopher David Hume encapsulated a sanguine view of British imperial activity in this period, depicting it as a mutually advantageous undertaking. He claimed that, whereas the Spanish Empire had arisen through a noxious mixture of 'sloth ... avidity and barbarity', British colonies were 'established on the noblest footing' by 'industrious planter[s]' whose labour 'promoted the navigation, encouraged the industry, and even perhaps multiplied the inhabitants of their mother-country', so that 'the spirit of independency, which was reviving in England, here shone forth in its full lustre'.[65] Reading this assessment from one of the great Enlightenment figures, the average cartographical enthusiast could be forgiven for thinking that to be on the receiving end of British imperialism was a stroke of good luck for a colony (even more so if the Spanish alternative was considered).

---

63. Pedley, *The Commerce of Cartography*, pp. 6-9.
64. *The Nautical Almanac and Astronomical Ephemeris, for the Year 1767* (London: W. Richardson & S. Clark, 1766); H. Overton, *A New & Correct Map of the Trading Part of the West Indies, including the Seat of War between Gr. Britain and Spain: Likewise the British Empire in America, with the French and Spanish Settlements Adjacent* (London: Henry Overton, at the White Horse without Newgate, 1741); M.K. Barritt, 'Navigational Enterprises in Europe and Its Empires, 1730-1850', *The Mariner's Mirror* 102, no. 2 (2016), 231-233.
65. D. Hume, *The History of England from the Invasion of Julius Caesar to the Revolution in 1688* (London: J. Mcreery, 1807), Vol. 6, p. 186.

However, the form of imperial activity that Hume upheld as a sort of archetype of benign international engagement was already giving way by this time to a much more bellicose interventionist form. The year 1772 was a turning point in this respect. That year the British East India Company's presence in Bengal shifted from being commercial to colonial. The rules of trade eventually required the rule of (British) law to uphold them and, in the process, the way the world was seen by Britain altered. As Warren Hastings, Bengal's new governor, pointed out the following year, '[a] system of affairs so new requires a new system of government to conduct it'[66] and that is precisely what happened. The empire of trade rapidly expanded into the business of colonial rule, territorial occupation and cultural transformation, with all the chaos, distress and factious change that followed in its wake.

Maps continued to be published as the character of the fledgling British Empire evolved, reflecting as best as they could the developments in the ways that the world was seen politically and culturally, as well as geographically. Sometimes, cartographers led the way, incorporating all the latest intelligence into their works. Other times, though, they lagged behind. Probably the most conspicuous example of the latter occurred around 1744 – a full century after Europe became aware of the location of New Zealand's coast. This was when the English mapmaker Richard Seale produced *A Chart Shewing the Track of the* Centurion *Round the World*,[67] which was intended to form the insert of a book published in 1748 detailing the circumnavigation of the world by HMS *Centurion* between 1740 and 1744 – a disaster-plagued expedition led by Commodore George Anson.[68]

Seale conscientiously plotted the course in his world map that Anson's fleet had followed. However, New Zealand was missing from the chart altogether (as was Tasmania and part of the Australian coast). There are various possible explanations for this omission, but it does hint that New Zealand's existence was far from a major focus for British readers – a point that was reinforced by the fact that this error was not remedied in the subsequent fourteen editions of Anson's book that were

66. W. Hastings to G. Colebrooke, 7 March 1773, in G.R. Gleig (ed.), *Memoirs of the Life of the Right Hon. Warren Hastings* (London: R. Bentley, 1841), Vol. 1, pp. 289-90.
67. Richard Seale, *A Chart Shewing the Track of the* Centurion *Round the World* (London, *c*. 1744), National Library of Australia, Rex Nan Kivell Map Collection, Call number MAP NK 4592.
68. G. Anson, *A Voyage Round the World in the Years* MDCCXL, I, II, III, IV (London: John and Paul Knapton, 1748), insert.

published over the following quarter of a century.[69] Incidentally, four years after he produced this chart, Seale went on to design a map for the Hudson's Bay Company, showing alternative routes to the Pacific. It was regarded as so 'bizarre' by the company's board of governors that they insisted that all copies of it be destroyed.[70] Seale's apparent laxity serves as a reminder that cartographical knowledge in this era was still inaccurately reproduced on occasion and that the means of cross-refencing information contained in maps was not especially thorough, and could not always keep pace with the rate of publication.

New Zealand was returned to the world in 1766 in a map that was 'Engraved by the Kings Authority for the New Geographical Dictionary'.[71] The creator of this work was named as Prinald, about whom practically nothing is known. The map itself was hand-coloured, with only some of the nations labelled (including 'New Zeeland'), but no locations within any territory were identified. Two years later, John Callander's *Reduced Chart of Australasia*, appeared in his book *Terra Australis Cognita, or, Voyages to the Terra Australis.*[72] It contained the outline of New Zealand (which he spelt 'New Zelande') and its shape resembled those of the territory in previous maps (excluding, of course, Bellin's aberration fifteen years earlier). However, while the major place-names on the work – such as 'Eastern Ocean', 'New Holland' and 'South Sea' – were in English, the locations within the territories that appeared on the chart were almost all in French, which was as good a clue as any as to the origins of the map that Callander based his work on (de Vaugondy's *Carte réduite de l'Australasie*). It was also an indication that Callander felt no need to anglicise such information for his audience, which reflected the inattentiveness he gave to the work as a whole, extending from its diminished cartographical detail to its roughly etched text. This map was very much in keeping with the type that was used as an appendage to a book, rather than a display item in its

69. As an example, see G. Anson, *A Voyage Round the World in the Years* MDCCXL, I, II, III, IV (London: H. Woodfall, 1769), insert.
70. B. Warren, 'Maps as Social History (Review)', *Huntington Library Quarterly* 64, nos 1/2 (2001), pp. 275-76.
71. Prinald, *A New Map of the World on Mercator's Projection* (1766), National Library of Australia, Bib ID 1773664.
72. J. Callander, 'Reduced Chart of Australasia', in J. Callander and C. de Brosses, *Terra Australis Cognita, or, Voyages to the Terra Australis, or Southern Hemisphere, During the Sixteenth, Seventeenth, and Eighteenth Centuries* (Edinburgh: J. Donaldson, 1768), Vol. 3, insert.

own right, which was the more popular form in this era (and the more lucrative for mapmakers).

* * *

For the most part, British-produced maps of the world made during the 1750s and 1760s had all settled into a relatively harmonious consensus on the shape and location of New Zealand. There were no vigorous debates about whether there was a channel or strait between its northern and southern regions, no one appeared interested in ascertaining the territory's exact latitude, and there was little concern over the conflicting names given to portions of its coastline. New Zealand was a distant hint of land that for most cartographers was simply not worth excessive attention.

Ironically, though, what did preoccupy the thoughts of a number of mapmakers at this time was not the presence of a particular territory but the absence of one – specifically, Terra Australis Incognita, the unknown southern land. This long-held belief stemmed from an intuitive conviction that the world's form had to be roughly proportionate, and so the preponderance of continental territory in the northern hemisphere ought to have its counterpoint somewhere in the antipodes.[73] The mystery of the possible existence of this Great Southern Continent – a landmass that lay frustratingly just beyond the reach of previous European expeditions – only added to its enticement.

One of the last serious advocates of the existence of the Great Southern Continent in this era was the Scottish hydrographer, Alexander Dalrymple.[74] As an official in the British East India Company in Madras in the 1750s, Dalrymple had access to the company's maps and records, which captivated him to such an extent that he rejected a promotion in favour of pursuing his bourgeoning interest in geography and exploration. Between 1759 and 1764, Dalrymple undertook three exploratory voyages through various parts of Southeast Asia and, on the basis of the work he undertook in these expeditions, he campaigned, with the support of the Royal Society, to lead a planned voyage to the South Pacific later in the decade.[75] His personal skills, though, were on occasion subject

73. B. Douglas, '"Terra Australis" to Oceania: Racial Geography in the "Fifth Part of the World"', *Journal of Pacific History* 45, no. 2 (2010), 179-210.
74. D.A. Lanegran, 'Alexander Dalrymple: Hydrographer' (PhD thesis, University of Minnesota, 1970), p. 12.
75. A.S. Cook, 'Alexander Dalrymple (1737-1808), hydrographer to the East India Company and the Admiralty, as publisher: a catalogue of books and charts (PhD thesis, University of St Andrews, 1993), pp. 18-21.

to question, with one journal observing that '[h]owever equal this gentleman maybe to the execution he appears to have but an indifferent talent at negociation'.[76]

From the mid-1760s, Dalrymple adopted the well-worn theory that the world's landmass needed to be balanced in order for it to rotate as it did and he augmented this idea by claiming that specific wind patterns in the southern hemisphere were the result of a large continent that lay somewhere at a latitude further south than had yet been explored. 'It cannot be doubted from so many concurrent testimonies', he wrote assuredly, 'that the SOUTHERN CONTINENT has been already *discovered* on the east side; and it appears more than probable, that TASMAN'S discovery, which he named STAAT'S LAND, but which the maps call NEW ZEALAND, is the *western* coast of this *Continent*.'[77] However, in the absence of firm cartographical evidence collected from any previous European visitors to the region, Dalrymple's map of New Zealand, which he completed sometime in the late 1760s, was almost a replica of the chart that Visscher had produced of the territory.[78]

By contrast, in 1763, *The Gentleman's Magazine* published a map of New Zealand which clearly showed the territory as an outcrop of the Great Southern Continent. The map had been created by the French cartographer, Philippe Buache, and was now being introduced to British readers. Since 1752, Buache had concerned himself with the form of the Antarctic region and a decade later had concluded that '*New Zeland*, which was discovered by a *Dutchman*, one *Tasman*, in 1642 … [is] a very high promontory [and] appeared to me to be the head of that Antarctic chain of mountains'. He conceded that this was a theoretical impression of the region[79] but, to readers of the magazine, there was a certain logic to the depiction of New Zealand as part of a continent rather than an insular territory.

* * *

76. *The Monthly Review; or, Literary Journal* 40 (London: R. Griffiths, 1769), p. 94.
77. A. Dalrymple, *An Historical Collection of the Several Voyages and Discoveries in the South Pacific Ocean*, 2 vols (London: J. Nourse, 1770-71), Vol. 2, p. 22.
78. Ibid., Vol. 1, insert. Note, the date of this map is 1769; ibid., Vol. 2, insert. B. Hooker, 'James Cook's Secret Search in 1769', *The Mariner's Mirror* 87, no. 3 (2001), p. 298.
79. *The Gentleman's Magazine* 33 (London, 1763), pp. 32-33. Also see V. Collingridge, 'Mapping the Fantastic Great Southern Continent, 1760-1777: A Study in Enlightenment Geography', *The Cartographic Journal* 57, no. 4 (2020), p. 342.

When the possibility of a British expedition to observe the transit of Venus in the southern hemisphere began to materialise in 1767, the Royal Society convened a special committee of eight fellows, presided over by the Astronomer Royal, to determine who might lead it. After a short deliberation, the committee unanimously recommended Dalrymple 'as a proper person to send to the South Seas, having a particular turn for discoveries, and being an able navigator and well skilled in observation'. The Royal Society endorsed the committee's recommendation and forwarded Dalrymple's name to the relevant authorities as the expedition's preferred leader.[80]

If Dalrymple's theory about New Zealand being part of the Great Southern Continent was true, this would be a momentous opportunity to travel to the region and prove it. One of the great geographical puzzles that had led to centuries of conjecture looked like it was about to be resolved and would lead to a cartographical revolution – a substantially new depiction of the southern hemisphere, based on scientific certainty, not speculation. The vast landmass which Dalrymple was convinced lay in the region would finally be located, claimed, named and mapped and, in the process, the perimeter of Europe's knowledge of the world would be extended enormously.

80. K. Tregonning, 'Alexander Dalrymple: The Man Whom Cook Replaced', *Australian Quarterly* 23, no. 3 (1951), p. 60.

Chapter 7

'To Add a Lustre to this Nation': Cook's Expedition

Readying the *Endeavour* for the Southern Hemisphere

What better way to arouse popular enthusiasm for a project than with a celestial visitation? That is precisely what the Royal Society did when it began to consider how it would go about obtaining government backing for its ambitious plans to observe the transit of Venus – the passage of that planet between the sun and earth – which was next due to take place in 1769. True to its reputation, the English weather had obstructed a view of the previous transit in 1761, which led the Oxford astronomer Thomas Hornsby to propose a more far-reaching effort for the next appearance of this phenomenon. In December 1765, he submitted a paper on the topic to the Royal Society, which by this time was one of Europe's foremost scientific bodies, and which was enjoying an increasingly intimate relationship with the British government.[1]

Hornsby furnished the Royal Society with a wealth of coordinates, tables and calculations to support his case but, for those readers whose specialty areas lay elsewhere, it was his concluding comments which outlined the historic magnitude that the event would have, not just for the scientific community, but for the honour of the nation. 'An opportunity of observing another transit of Venus will not again offer itself till the year

1. H.G. Lyons, *The Royal Society, 1660-1940: A History of Its Administration under Its Charters* (Cambridge: Cambridge University Press, 1944), pp. 159-209.

1874', he informed the Royal Society's Council and Fellows: 'It behoves us therefore to profit as much as possible by the favourable situation of Venus in 1769.' The likely competitive stance of other countries, Hornsby believed, placed a particular burden on Britain to take the lead in this great scientific enterprise:

> We may be assured that several Powers of Europe will again contend which of them shall be most instrumental in contributing to the solution of this grand problem. Posterity must reflect with infinite regret upon their negligence or remissness; because the loss cannot be repaired by the united efforts of industry, genius, or power. How far it may be an object of attention to a commercial nation to make a settlement in the great Pacific Ocean, all to send out some ships of force with the glorious and honourable view of discovering lands towards the South pole, is not my business to enquire. Such enterprizes, if speedily undertaken, might fortunately give an advantageous position to the astronomer, and add a lustre to this nation, already so eminently distinguished both in arts and arms.[2]

Observing a planetary movement had somehow veered into the very terrestrial realm of imperialist ambition and, despite Hornsby's evasive claim that this was none of his business, the fact was that science and Empire were unavoidably entangled in this proposed venture. The Royal Society was the most successful example in this period of the sciences coalescing around a single institution. It was also the institution through which royal grants for various scientific undertakings were funded. In the second half of the eighteenth century, the British government's growing concern with scientific matters dovetailed very comfortably with the developing purpose and functions of the Royal Society.[3] The society served as a deep well of expertise that the government could draw on whenever it needed to and, in return, officials could ensure that it remained appropriately funded, at least for specific joint projects.

2. T. Hornsby, 'XXXIV. On the transit of Venus in 1769. To the Right Honourable The Earl of Morton, President to the Council and Fellows of the Royal Society, this discourse is, with all humility, inscribed, by their humble servant, Thomas Hornsby', *Philosophical Transactions* 55 (1765), pp. 343-44.
3. Gascoigne, 'The Royal Society and the Emergence of Science as an Instrument of State Policy', p. 177.

As part of the (costly) global campaign to observe the transit of Venus, an observation post somewhere in the South Pacific would be needed. It was not just the British who were aware of this, though. Throughout Europe, scientists were mobilising to pool their knowledge, aware that this would be a once-in-a-lifetime opportunity to glean information about this planetary event. In 1762, the French astronomer Jérôme Lalande distributed copies of his *Mappemonde* to colleagues across the continent.[4] This map (which included a rough representation of New Zealand's west coast) projected the locations around the world where the transit of Venus could best be observed, including an elliptical line curving near Tahiti.

Armed with all this information, and acutely aware of the importance of this forthcoming astronomical event, in February 1768, the Royal Society agreed to approach King George III for financial support for an audacious expedition which would include a journey into stretches of the South Pacific that were still barely known to Europe. The handwritten memorial to the king began with an explanation of what the transit of Venus was and noted that several other European nations were well advanced in their own plans to observe it. This was an appeal to patriotism as much as to the scientific importance of the event. '[T]he British nation have been justly celebrated in the learned world for their knowledge of Astronomy, In which they are inferior to no nation upon earth', the memorialists advised their monarch, before warning that 'it would cast dishonour upon them should they neglect to have correct observations made of this important phenomenon'. A spring departure of observers from Britain to the southern hemisphere was proposed, and an estimate of the costs for the expedition provided:

> the expense of having the Observations made in the places above specified, including a reasonable gratification to the persons employed, and finishing them with such instruments as are still wanting, would amount to about 4000 pounds, exclusive of the expense of the ships which must convey in return the observers that are to be sent to the Southward of the Equinoxical line.[5]

4. J. Lalande, *Mappemonde* (Paris, 1762), in R. Woolley, 'Captain Cook and the Transit of Venus of 1769', *Notes and Records of the Royal Society of London* 24, no. 1 (1969), p. 22.
5. Weld, *A History of the Royal Society, with Memoirs of the Presidents*, Vol. 2, pp. 33-38.

The Royal Society made it clear that it had no funds at all for such a venture (and went as far as to claim that it was barely able even to cover its own operating costs). The response from George III was almost immediate, which was possibly an indication of his personal enthusiasm for the proposed expedition. On 24 March the society's president announced to its members that the entire sum that it had sought had been paid by the Treasury, with any surplus to be credited to the society.[6]

Having been victorious over the French in the Seven Years War in 1763 and having enlarged its navy, grown its interests in America and secured Canada as a colony, the potential that an expedition into the South Pacific opened up was too alluring to pass over. The transit of Venus would serve as a convenient pretext; it would allow Britain to extend its presence in that part of the world, hopefully without generating undue suspicion among the French or Spanish. The fact that it was a valid and important scientific undertaking certainly helped in this objective. The expedition would thus kill two birds with one stone: enabling Britain to further its interests in a region where Europe had yet to assert its presence in any significant way; and contributing to the advance of scientific knowledge. With the king supporting and financing the society's proposed expedition, it was now up to the Admiralty to allocate a ship and crew for the venture,[7] with the society responsible for hiring astronomical observers to accompany it.

The name that kept surfacing in discussions over who should lead this expedition was Dalrymple. He was 'thought of as a proper person to be employed on that service, and for prosecuting discoveries in that quarter [the South Pacific]'; and he even accompanied the Surveyor of the Navy to examine two vessels that might be used for the expedition (and that were subsequently purchased).[8] Unsurprisingly though, a role as coveted as this inevitably led to a good deal of surreptitious jostling among those who looked at this venture as an opportunity for furthering their own particular ends. Just when it seemed that Dalrymple's appointment to lead the expedition was little more than a formality,[9] a sudden power struggle broke out between the Royal Navy and the Royal Society over

6. Ibid.

7. L. Withey, *Voyages of Discovery: Captain Cook and the Exploration of the Pacific* (Berkeley: University of California Press, 1989), pp. 18-21.

8. Anonymous, 'Biographical Memoir of Alexander Dalrymple, Esq., Late Hydrographer to the Admiralty', *The Naval Chronicle* 35 (1816), pp. 194-95.

9. A. Dalrymple, *An Account of What Has Passed between the India Directors and Alexander Dalrymple* (London: J. Nourse, 1769), pp. 16 ff.; Anonymous,

control of the expedition. Admiral Edward Hawke, the First Lord of the Admiralty, threw a fly in the ointment by insisting that a naval officer be appointed to head the voyage, otherwise he could be liable to parliamentary impeachment.[10] A compromise was proposed by Hawke, whereby Dalrymple would be put in charge of the scientific aspects of the expedition, but Dalrymple saw this as a form of divided command which could hamper the entire undertaking. In an act of career sacrifice (however chivalrous it might have appeared), Dalrymple stepped away from any subsequent involvement in the planned voyage to the South Pacific.[11]

With the loss of Dalrymple, the need for the Royal Society to find a replacement astronomer became pressing. Attention turned briefly to Oxford and Cambridge universities but this yielded no prospective candidates to command the expedition. The Reverend Nevil Maskelyne, a Fellow of the Royal Society and the Astronomer Royal was approached but expressed no interest. He suggested the astronomer William Wales but, by this time, he had already made arrangements to travel to Canada to observe the transit of Venus there. However, while the attention of the Royal Society was focussed on finding a suitable astronomer, the Admiralty had been carrying out its own search for someone to command the expedition and, in early April, it formally appointed Captain James Cook to the role.[12]

Cook would be joined on the voyage on HMS *Endeavour* by a group of scientists to fulfil the ambitions of the Royal Society. These included Charles Green, who was assistant to the Astronomer Royal and who had the backing of the Royal Society, and a group led by Joseph Banks (who financed his and his team's own passage to the tune of £10,000 – more than double what George III contributed to the entire expedition). Included in Banks' party were the naturalist Daniel Solander, the draftsman Herman Spöring and two artists, Alexander Buchan and Sydney Parkinson. No previous expedition to the region had anything like this level of expertise and, from a cartographical perspective, the

'Memoirs of Alexander Dalrymple Esq.', *The European Magazine and London Review* 42 (November 1802), p. 325.

10. Ibid.
11. H.T. Fry, 'Alexander Dalrymple and Captain Cook: The Creative Interplay of Two Careers', in R. Fisher and H. Johnston (eds), *Captain James Cook and His Times* (Canberra: Australian National University Press, 1979), pp. 46-47.
12. Withey, *Voyages of Discovery*, pp. 18-21.

potential for whatever maps Cook produced to be fleshed out with all forms of supplementary scientific information was unprecedented.

Cook's expedition would also be furnished with the very latest in navigational equipment, which would enable him to produce the most accurate charts possible at the time. Mapmaking by the late eighteenth century had become more complex and involved several disciplines, including conventional cartography, as well as topography, surveying, chorography, geodesy and navigation.[13] The apparatus that was loaded onto the *Endeavour* gives some idea of how far mapmaking had advanced since the last European visit to New Zealand. The master astronomical instrument-maker, John Bird, supplied Cook with a twelve-inch quadrant on a pillar stand, with two plumb bobs, two glass beakers and two detachable reading microscopes.[14] Such a device could be used to help to fix a ship's position and establish latitude with considerable accuracy.[15] There were also two Gregorian reflecting telescopes,[16] which utilised a pair of concave mirrors to reflect images to an eyepiece, and which gave much better views of terrestrial objects than the less sophisticated telescope Tasman had relied on.[17]

Crucially, the expedition had access to one of the most important technological developments for navigation up until this time: an advanced sextant (in this case, a Ramsden sextant), which made it possible for navigators to establish latitude to between one and two nautical miles.[18]

---

13. R. Sorrenson, 'The Ship as a Scientific Instrument in the Eighteenth Century', *Osiris* 11 (1996), p. 223.
14. Twelve-inch quadrant by John Bird (London, 1760-69), Science Museum Group, Object Number 1876-572/1.
15. A. Chapman, 'The Accuracy of Angular Measuring Instruments Used in Astronomy between 1500 and 1850', *Journal for the History of Astronomy* 14, no. 2 (1983), p. 135.
16. C.E. Herdendorf, 'Captain James Cook and the Transits of Mercury and Venus', *Journal of Pacific History* 21, no. 1 (1986), p. 46.
17. N. Maskelyne, 'XLIX. Description of a Method of Measuring Differences of Right Ascension and Declination, with Dollond's Micrometer, together with Other New Applications of the Same', *Philosophical Transactions* 61 (1771), 536-46.
18. A. Stimson, 'The Influence of the Royal Observatory at Greenwich upon the Design of 17th and 18th Century Angle-Measuring Instruments at Sea', *Vistas in Astronomy* 20 (1976), pp. 123-24; N.A. Doe, 'The Potential Accuracy of the Eighteenth-Century Method of Determining Longitude at Sea', *SILT – A Journal of Personal Research* 556 (2016), p. 2.

This revolutionised cartography, making it possible to map territories and routes with far more precision than had previously been possible.

There were also barometers, alarum clocks, additional sextants, thermometers, compasses, theodolites, Gunter's chains (for measuring distances on land) and dipping needles[19] (magnetised needles used to make measurements connected with compass directions),[20] all of which had to be housed on the ship along with provisions for the 85 men that were crammed on board this 30-metre vessel. Cook and Green later reported back to the Royal Society on how some of this equipment was used during the expedition:

> The astronomical clock, made by Shelton and furnished with a gridiron pendulum, was set up in the middle of one end of a large tent, in a frame of wood made for the purpose at Greenwich, fixed firm and as low in the ground as the door of the clock-case would admit, and to prevent its being disturbed by any accident, another framing of wood was made round this, at the distance of one foot from it. The pendulum was adjusted exactly to the same length as it had been at Greenwich. Without the end of the tent facing the clock, and 12 feet from it, stood the observatory, in which were set up the journeyman clock and astronomical quadrant: this last, made by Mr. Bird, of one foot radius, stood upon the head of a large cask fixed firm in the ground, and well filled with wett heavy sand. A centinel was placed continually over the tent and observatory, with orders to suffer no one to enter either the one or the other, but those whose business it was. The telescopes made use of in the observations were — Two reflecting ones of two feet focus each, made by the late Mr. James Short, one of which was furnished with an object glass micrometer.[21]

---

19. I. Kaye, 'Captain James Cook and the Royal Society', *Notes and Records of the Royal Society of London* 24, no. 1 (1969), p. 8.
20. G. Graham, 'Observations of the Dipping Needle, Made at London, in the Beginning of the Year 1723. By Mr. George Graham, Watchmaker, FRS', *Philosophical Transactions (1683-1775)* 33 (1724), 332-39.
21. J. Cook and C. Green, 'Observations Made, by Appointment of the Royal Society, at King George's Island in the South Sea; By Mr. Charles Green, Formerly Assistant at the Royal Observatory at Greenwich, and Lieut. James Cook, of His Majesty's Ship the Endeavour', *Philosophical Transactions (1683-1775)* 61 (1771), pp. 397-98.

Cook lauded the 'Mathematical Instrument makers for the improvements and accuracy with which they make their Instruments, for without good Instruments the Tables would loose part of their use' and, in relation to the sextants in particular, he noted that 'we cannot have a greater proof of the accuracy of different Instruments than the near agreement of the above observations, taken by four different Sextants and which were made by three different persons'.[22]

There was one other item that Cook depended on greatly in his navigation and mapmaking during this expedition: Maskelyne's *Nautical Almanac*.[23] This work, published in 1766, provided detailed coordinates for the positions of various celestial bodies throughout the year, which represented a significant advance for calculating longitude for ships while at sea. Ascertaining longitude had bedevilled sailors since they had begun to embark on long-distance voyaging (unlike latitude, which could be determined by observing the altitude of the sun). It had become so acutely important for nations engaged in oceanic travels that the British parliament passed the Longitude Act 1714 by which anyone who could devise a means of determining longitude at sea would be awarded prize money. Proposals poured in, many of them ranging from dubious to absurd, but such was the navigational priority of being able to establish longitude that the quest continued.[24] In 1752, the London periodical, *The Gentleman's Magazine*, which also sought suggestions for resolving this puzzle, informed its readers that 'many schemes have been sent us, which to publish would do no honour to their authors nor service to the community'.[25] Hundreds of plans and formulae were devised but most missed the mark and were politely rejected.

The breakthrough came in 1755, when the German astronomer Tobias Mayer devised a set of lunar tables that could be used for predicting the position of the moon at any given time, thus enabling longitude to be determined by what became known as the lunar distance method

22. J. Cook, in E.G.R. Taylor, *The Mathematical Practitioners of Hanoverian England, 1714-1840* (Cambridge: Cambridge University Press, 1966), p. 202.
23. N. Maskelyne, *The Nautical Almanac and Astronomical Ephemeris* (London: J. Nourse, 1766); Sorrenson, 'The Ship as a Scientific Instrument in the Eighteenth Century', p. 225. E. Forbes, *Greenwich Observatory: The Royal Observatory at Greenwich and Herstmonceux: Volume 1: Origins and Early History (1675-1835)* (London: Taylor & Francis, 1975), p. 22.
24. A.J. Kuhn, 'Dr. Johnson, Zachariah Williams, and the Eighteenth-Century Search for the Longitude', *Modern Philology* 82, no. 1 (1984), p. 40.
25. *The Gentleman's Magazine* 22 (London, August 1752), p. 359.

(which involves calculating the distance between the moon and the sun or a fixed star). These tables were analysed at the Greenwich Royal Observatory and found to be sufficiently accurate that they were published in English, with additional calculations and guidance on how to use the tables for navigation.[26]

With this cluster of materials and equipment safely stowed below deck, Cook was about to command an expedition that would venture to where all the coordinates and territories on Europe's collective corpus of maps vanished. Science could only take him so far. Where the charts went blank, he would have to depend on his perception and intuition, along with the navigational skills that he had acquired so far, to plot new courses and outline new territories. However, with exploration came obligation. This expedition would not be some act of reckless speculation for cheap glory. Serious money and even more serious scientific thought had been invested in Cook's voyage and the British government and the Royal Society rightly expected serious dividends in return.

Cook Leaves Plymouth

Although in the preceding 120 years, Europe's reading classes had enjoyed access to a profusion of maps which in one way or another depicted New Zealand on them, these were all based on the charts produced during Tasman's expedition. The Dutch explorer's outline of the territory was the common ancestor of all subsequent impressions of what by the 1760s in Britain had become known as New Zealand. Moreover, still in this decade, for some (including the Admiralty) that incomplete line etched in the South Pacific hinted at the possible existence of the Great Southern Continent. Dalrymple, who had been a contender to command the expedition that Cook was now in charge of, remained adamant about the existence of this super-continent. Logic determined it, mathematics enabled its scale to be calculated and an imperial imagination conceived its prospects:

> The number of inhabitants of the Southern Continent is probably more than 50 millions, considering the extent of the eastern parts discovered by Juan Fernandez, and to the

---

26. M. Croarken, 'Providing Longitude for All', *Journal for Maritime Research* 4, no. 1 (2002), pp. 107-8; G. Dolan, *The Quest for Longitude and the Rise of Greenwich: A Brief History* (London: The Royal Observatory Greenwich, 2022).

> western coast seen by Tasman ... [which] amounts to 4596 geographic miles. This is greater extent than the whole civilised part of Asia, from Turkey, to the eastern extremity of China. There is at present no trade from Europe thither, though the scraps from this table would be sufficient to maintain the power, dominion, and sovereignty of Britain, by employing all its manufactures and ships.[27]

Cook's ability to use the latest range of navigational instruments at his disposal was probably unmatched in any other individual in Britain at this time. The fact that there were expert astronomers travelling with him only enhanced the intelligence-gathering possibilities of this expedition. However, technical expertise with this equipment, along with detailed data on celestial movements, were only part of what was required if Cook was to map successfully the furthest reaches of the world. Navigation still depended on the tough lessons of experience as much as information acquired from books. Grappling with ice, reefs, sandbars and the wild, terrifying immensity of the still little-known South Pacific Ocean would be challenges for which instinct and understanding would be vital.[28]

Cook's training ground had been along the wind-battered coast of Yorkshire, where sediment-lined shallows often posed as much of a risk as rocky outcrops and where an awareness of the risks of currents and tides was acquired through keen observation and maintained in the memory. Such skills made Cook useful to the Navy and, in the 1760s, having been promoted to the rank of master, he employed them when undertaking hydrographic and cartographical work along the Newfoundland coast, producing charts that became an archetype of mapmaking precision in this period.[29] In writing to John Clevland, the Secretary to the Admiralty, in December 1762, Admiral Lord Colville

27. Dalrymple, *An Historical Collection of the Several Voyages*, Vol. 1, pp. xxvii-xxix,
28. J.C. Beaglehole, 'Cook the Navigator' (A lecture delivered to the Royal Society on 3 June 1969 on the occasion of the celebration of the observation of the transit of Venus by Captain James Cook, R.N., F.R.S.), *Proceedings of the Royal Society of London (Series A, Mathematical and Physical Sciences)* 314, no. 1516 (1969), p. 28.
29. O.U. Janzen, 'The Making of a Maritime Explorer: James Cook in Newfoundland, 1762-1767', *The Northern Mariner/Le marin du nord* 28, no. 1 (2018), 23-38; J. Cook, *Directions for Navigating the West-Coast of Newfoundland, with a Chart Thereof* (London: J. Mount & T. Page, 1766).

mentioned that Cook had recently furnished him with drafts of the charts he had been preparing:

> I beg leave to inform their Lordships that, from my experience of Mr. Cook's genius and capacity, I think him well qualified for the work he has performed, and for greater undertakings of the same kind. These draughts being made under my own eye, I can venture to say they may be the means of directing many in the right way, but cannot mislead any.[30]

Such fulsome praise from such a senior officer would not go unnoticed among the decision-makers in the Navy.

Cook returned to Britain from Newfoundland in November 1767 and, just six months later, the Navy commissioned him for the ambitious new expedition to the southern hemisphere. In preparation for this voyage, Cook began to seek out any navigational material that would be of use, which included a published version of Tasman's records. He also drew on more current cartographical intelligence on the region assembled by Samuel Wallis, who had recently returned to England from a journey during which he had claimed Tahiti for Britain (and which he named King George III Island).[31] Cook was given a copy of the details of Wallis' expedition, including the charts that he had produced.[32] In addition to these sources, Banks also passed on to Cook a copy of a short, unpublished work by Dalrymple, which contained cartographical intelligence that the British had obtained from Manila during their defeat of Spanish forces there in 1762.[33] Cook assessed this mass of materials closely and, from it, extracted the details he felt would be necessary for his own expedition. As he was venturing into unfamiliar regions, every piece of cartographical information took on a heightened value to him.

---

30. A. Colville to J. Clevland, 30 December 1762, in *Historical Records of New South Wales: Volume 1, Part 1: Cook 1762-1780* (Sydney: Charles Potter, 1893), p. 299.
31. A. G. Price, ed., *The Explorations of Captain James Cook in the Pacific* (New York: Dover Publications, 1971), 17; R. Wilson, *Voyages of Discoveries Round the World* vol. 1 (London: James Cundee, 1806), 168-9.
32. S. Wallis, 'Log books and sketchbook kept during his voyage around the world in command of H.M.S. Dolphin', British Admiralty (Adm. 55/35, PRO Reel 1579, in National Library of New Zealand, Micro-MS-0353.
33. H.T. Fry, 'Alexander Dalrymple and New Guinea', *Journal of Pacific History* 4, no. 1 (1969), p. 86; C. Markham (ed.), *The Voyages of Pedro Fernandez de Quiros, 1595-1606*, 2 vols (London: Hakluyt Society, 1904), Vol. 2, pp. 517-36.

Had this been a straightforward voyage to Tahiti to observe the transit of Venus, then many of the maps and documents that Cook gathered would have been surplus to requirements (particularly Tasman's charts, which would only have been of peripheral value). However, the Admiralty had more far-reaching plans than simply sending a ship to almost the other side of the world to observe a planetary movement. On 30 July 1768 – just 27 days before Cook was due to board the *Endeavour* to commence his voyage – he was handed a sealed envelope with 'Secret. Additional Instructions for Lt James Cook' written on it, which laid out what was expected of him once he had concluded recording the transit of Venus. Despite the supposed secrecy of these orders, though, Cook would have at least had some inkling that such a well-funded and well-equipped expedition had not been organised merely to witness a celestial event that in practical terms was of little material value. Both the scale of preparations and the fact that he had gone to considerable lengths to acquire seemingly irrelevant cartographical intelligence hint that Cook had at the very least some vague foreknowledge of what the Admiralty had in mind for this voyage.[34]

There is no record of exactly when Cook opened these confidential instructions but the audacity of aspiration that they represented would have startled even someone of his experience and enterprise. They commenced with a reference to the need to uphold Britain's honour as a maritime power and the need to advance trade and navigation as being the overriding bases for this additional purpose of the expedition. The Admiralty believed that there was 'reason to imagine that a continent, or land of great extent, may be found to the southward of [Tahiti]' and so Cook was instructed to sail to the latitude of 40 degrees south (which roughly bisects New Zealand's North Island) 'until you discover it or fall in with the Eastern side of the land discovered by Tasman and now called New Zealand'. If the great continent was discovered, Cook was required to chart its location and provide details on its geography and the sea movements around its coast.

Cook was also instructed to land on any territory he encountered, and obtain details about its geology, fauna and flora (including collecting samples of plants and seeds where possible). Moreover, presuming that some of these territories would be inhabited, the Admiralty outlined its expectations for Cook:

> You are likewise to observe the genius, temper, disposition and number of the natives, if there be any, and endeavour by

34. Hooker, 'James Cook's Secret Search in 1769', p. 299.

> all proper means to cultivate a friendship and alliance with them, making them presents of such trifles as they may value, inviting them to traffic. … You are also with the consent of the natives to take possession of convenient situations in the country, in the name of the King of Great Britain; or, if you find the country uninhabited, take possession for His Majesty by setting up proper marks and inscriptions, as first discoverers and possessors.[35]

On the matter of New Zealand, the British were obviously eager to resolve the puzzle of the territory's shape. On reaching its shores, Cook was ordered to observe with care:

> the latitude and longitude in which that land is situated, and explore as much of the coast as the condition of the Bark, the health of her crew, and the state of your provisions will admit of…. You will also observe with accuracy the situation of such islands as you may discover in the course of your voyage that have not hitherto been discovered by any Europeans, and take possession for His Majesty and make surveys and draughts of such of them as may appear to be of consequence, without suffering yourself however to be thereby diverted from the object which you are always to have in view, the discovery of the Southern Continent so often mentioned.[36]

The potential scale of intelligence that might be gathered as a result of carrying out these instructions was extensive, especially considering Cook's prodigious talents and the array of equipment and expertise on board the *Endeavour* to assist him. The Admiralty required that he send all his observations and findings to the Royal Society and, on returning home, he was to present all the logbooks and journals from the voyage to the Navy, which would keep them confidential until they decided exactly what to do with the expected wealth of detail that would be amassed during Cook's exploration.

---

35. Commissioners for executing the office of Lord High Admiral of Great Britain, 'Secret. Additional Instructions for Lt James Cook, Appointed to Command His Majesty's Bark the Endeavour', MS 2-Cook's voyage 1768-71 [manuscript]: copies of correspondence, etc., National Library of Australia, Bib ID: 1120886.
36. Ibid.

* * *

On 26 August 1768, the spectacle of departure played out at Plymouth, with Cook, the crew, assorted scientists and a year-and-a-half's provisions all packed tightly into the ship. This was the start of one of the greatest voyages of discovery in history, which over the following three years would radically reshape the impression of the world on maps – completing missing portions of existing territories and adding entirely new ones.

Five days earlier, a storm had moved into south-west England, with squally winds forcing everyone to seek shelter and preventing any vessels from entering or exiting the port. Eventually, conditions eased and the *Endeavour* set sail, with Banks barely able to suppress his exhilaration at the prospects of what lay ahead. 'After having waited in this place ten days', he wrote in his journal,

> the ship, and every thing belonging to me, being all that time in perfect readyness to sail at a moments warning, we at last got a fair wind, and this day at 3 O'Clock in the even weigd anchor, and set sail, all in excellent health and spirits perfectly prepard (in Mind at least) to undergo with Chearfullness any fatigues or dangers that may occur in our intended Voyage.[37]

The Outbound Journey

Within moments of being unfurled, the *Endeavour*'s sails became taut as they were filled with the easterly breeze streaming through the Plymouth dock. The first leg of the ship's voyage – to the island of Madeira (around 720 kilometres due west of Morocco) – was a well-traversed one, having been used by British vessels for around a century by this time.[38] Over the next several weeks, those on the ship who were unaccustomed to ocean voyaging began to adjust to life at sea. The wind always seemed either to be stiffening or slackening, the range of food available was narrow, topics of conversation thinned out over time, seasickness was frequent and the novelty of this form of travelling soon gave way to tedium.

37. J. Banks, *The Endeavour Journal of Joseph Banks, 1768-1771*, ed. J.C. Beaglehole, 2 vols (Sydney: Angus and Robertson Ltd, 1962), Vol. 1, p. 153 (hereafter referred to as Banks' Journal).
38. I.K. Steele, *The English Atlantic, 1675-1740: An Exploration of Communication and Community* (Oxford: Oxford University Press, 1986), p. 24.

Once out of sight of land, the immense scale of this watery world became fully apparent. Sea and sky joined in the haze of the horizon and, for those crammed on board the ship, the ocean must have felt boundless at times. The vessel was now utterly at the mercy of the elements. An early instance of this had occurred just six days after departing from Plymouth. A hard gale, accompanied by heavy rain, lashed the *Endeavour* for almost 24 hours. The iron futtock plate attached to the main topmast was ripped off in the wind and several dozen chickens were washed overboard, as was a small boat. Meanwhile, seawater was leaking through into some of the cabins,[39] giving everyone on board a rough reminder of the forces of nature that they would have to contend with on this journey.

After four days in Madeira, the next stretch along this aquatic highway was around 7,000 kilometres south across the Atlantic Ocean to Rio de Janeiro. During these two months of monotony, the trade winds took the *Endeavour* along the coast of West Africa before Cook steered south-west towards Rio de Janeiro, which the ship reached on 13 November. Every day, Cook was on deck, taking measurements to determine the vessel's location and calculating how far it had travelled. His observations were detailed, including cloud coverage, wind speed, sightings of land on the horizon, as well as the position of the sun, moon and stars, and sea-floor depths when approaching a coast. His entry for Friday 28 October is representative of the notes he kept as he guided the *Endeavour* to its next destination:

> Fresh Breeze and fine Clear weather. At a little past 1 a.m. Longitude in by the 3 following Observations – viz., by the Moon and the star Arietis, 32 degrees 27 minutes; by the Moon and Pollux, 32 degrees 0 minutes 15 seconds; by ditto, 31 degrees 48 minutes 32 seconds; the mean of the whole is 32 degrees 5 minutes 16 seconds West from Greenwich, which is 31 minutes more Westerly than the longitude by account carried on since the last Observation. The two first observations were made and computed by Mr. Green, and the last by myself. The star Arietis was on one side of the Moon

39. J. Cook, *The Journals of Captain James Cook on His Voyages of Discovery*, ed. J.C. Beaglehole (Cambridge: Hakluyt Society, 1955), Vol. 1, journal entry for 1 September 1768. Henceforth, references to the journals that Cook and others kept on the *Endeavour* voyage will be referred to by their author and the date of the journal entry.

> and Pollux on the other. This day at Noon, being nearly in the latitude of the Island Ferdinand Noronha, to the Westward of it by some Charts and to the Eastward by others, was in Expectation of seeing it or some of those Shoals that are laid down in most Charts between it and the Main; but we saw neither one nor a Nother. We certainly passed to the Eastward of the Island, and as to the Shoals, I don't think they Exhist [*sic*], grounding this my Opinion on the Journal of some East India Ships I have seen who were detain'd by Contrary winds between this Island and the Main, and being 5 or 6 Ships in Company, doubtless must have seen some of them did they lay as Marked in the Charts.* (* There is a very dangerous reef, As Rocas, 80 miles west of Fernando Noronha. The Endeavour passed 60 miles east of latter.) Wind South-East to South-East by East; course South 33 degrees West; distance 93 miles, latitude 3 degrees 41 minutes South, longitude 32 degrees 29 minutes West.[40]

For the previously landbound Banks, his observations were less mathematical and more environmental. At the same time that Cook was methodically working through the coordinates for the journey, Banks was detailing the experience:

> Almost immediately on crossing the tropick the air became sensibly much damper than usual, tho not materialy hotter, the thermometer then in general stood from 80 to 82. The nearer we approachd to the calms still the damper every thing grew, this was perceivable even to the human body and very much so, but more remarkably upon all kinds of furniture: every thing made of Iron rusted so fast that the knives in peoples pockets became almost useless and the razors in cases not free. All kinds of Leather became mouldy, Portfolios and truncks coverd with black leather were almost white, soon after this mould adheerd to almost every thing, all the books in my Library became mouldy so that they were obligd to be wiped to preserve them. ... The air during the whole time sin[c]e we crossed the tropick and indeed sometime before has been nearly of the same temperature throughout the 24 hours, the Thermometer seldom rising above a degree during

40. Cook's Journal, 28 October 1768.

> the time the sun is above the horizon. The windows of the cabbin have been open without once being shut ever since we left Madeira.[41]

The significance of these two extracts lies in how they reveal the complementary nature of the observations made by Cook and Banks. Both were specialists in their respective fields but, when the information that each man collected was combined, the resulting whole became greater than the sum of the parts.

Occasionally, details of the sources that were being relied on to navigate this route were revealed – usually inadvertently. One such case was Banks' passing comment about the winds as the *Endeavour* neared Rio de Janeiro, in which the botanist mentioned 'Dampier' – a reference to William Dampier, and his 1700 publication *A Collection of Voyages*, which contained a section on winds and currents in various parts of the world.[42] On another occasion, Cook referred to the copy of Tasman's journal he had, which was 'published by Dirk Rembrantse'. Actually, Cook was in possession of a condensed version of the journal which appeared (in an English translation) in John Narborough's 1694 book, *An Account of Several Late Voyages and Discoveries to the South and North*.[43] These were a sample of the sources that were used on Cook's voyage but, as with all the intelligence available, the art lay in selecting what was useful, discerning the degree of its reliability and knowing when to apply it. In this, Cook repeatedly demonstrated an almost instinctive mastery.

* * *

The extremely detailed record of coordinates and measurements that Cook was making as he sailed this route would assist enormously the ability of ships in the future to navigate across this portion of the Atlantic. It was the equivalent of providing comprehensive markings and signposting along a path where previous travellers had depended on a few rough traces and clues, and their own memories, to guide them through.

However, this was not some academic exercise conducted in the peaceful solitude of a ship's cabin. With almost 100 men crowded on

---

41. Bank's Journal, 25 October 1768.
42. W. Dampier, *Voyages and Descriptions* (London: James Knapton, 1700), Vol. 2.
43. J. Narborough, *An Account of Several Late Voyages and Discoveries to the South and North* (London: Sam Smith, 1694); Cook's Journal, 19 April 1770.

board the *Endeavour* for this journey, Cook had to uphold a high level of discipline if he was to accomplish what his instructions required. Months at sea in such a claustrophobic environment was bound to put a strain on everyone to some extent but it was the captain's responsibility to ensure that a minimum standard of order was maintained. Thus, when one sailor attempted to desert ship and another used abusive language to an officer, both were punished with twelve lashes, as was the boatswain's mate for failing to reprimand these two men sufficiently in the first place.[44] Thus, while charting was a crucial activity for Cook, he had numerous other responsibilities to manage simultaneously.

The *Endeavour*'s stay at Rio de Janeiro turned out to be unhappy for reasons other than sagging crew discipline. The Portuguese governor of the settlement was one of those officials who fluctuated from being pedantic to capricious – a man whom Banks waspishly described as 'illiterate' and 'impolite'.[45] Cook's patience was also strained times by having to deal with this bureaucrat, especially when he was forced to resort to written exchanges with the governor in order to make basic arrangements for the ship's supplies. The Portuguese were suspicious of Cook's aims, and with some basis as it turned out. The *Endeavour*'s captain made a detailed study of the military installations around the port and provided revised coordinates for the settlement's location for future use by the British Navy. Finally, though, with repairs to the ship complete, supplies replenished and the crew rested, on 7 December 1768, the *Endeavour* sailed out of Rio de Janeiro and, for the next six weeks, hugged the east coast of South America, until it sought shelter in a bay in Tierra del Fuego.

As ever, Cook was alert to the standard of maps he was using and the opportunity for improving them that his new navigational technology enabled. Delving into his prodigious knowledge of the route, he recalled that a Dutch squadron commanded by Admiral Jacques l'Hermite had visited this location in 1624 and mapped portions of the area. However, according to Cook:

> the account they have given of those parts is very short and imperfect, and that of Schouten and Le Maire still worse, that it is no wonder that the Charts hitherto published should be found incorrect, not only in laying down the Land, but in the Latitude and Longitude of the places they contain.

44. Cook's Journal, 30 November 1768.
45. Bank's Journal, 2 December 1768.

Then, in an unusually self-congratulatory tone, Cook boasted in his journal that: 'I can now venture to Assert that the Longitude of few parts of the World are better Ascertained than that of Strait Le Maire and Cape Horn, being determined by several Observations of the Sun and moon made both by myself and Mr. Green, the Astronomer.'[46] There is a building sense, as the voyage continued, that Cook was becoming ever more aware of the strength of his navigational prowess. He could not help noticing that literally every day, he was taking measurements and calculating coordinates that were adding substantially to the precision of charts of the routes he was sailing – something that his ability to determine longitude accurately was making possible in many cases for the first time.

By 26 January, the *Endeavour* had sailed round Cape Horn but, during the preceding day, Cook had not been sure of his exact location in relation to the Cape, such was the inadequacy of existing maps of the area that he was relying on. It did not help that a thick shroud of low-hanging cloud and fog had crept across this stretch of the South Atlantic, blotting out a view of the sun and coast, which Cook depended on for some of his measurements. He strained to establish whether the glimpses of land he saw were islands or part of the mainland and, in the end, gave up, describing it as being 'of very little Consequence to Navigation'.[47]

Over the next few days, Cook pushed southward, even though the conditions seemed to be conspiring against him. The ship was hit with gales, hail, rain and rough seas, but onward he went, plunging further south than any previous mariner of that time is known to have gone. On the final day of January 1769, having battered against worsening weather and plummeting temperatures over the previous day, and with the *Endeavour* advancing resolutely south into uncharted seas (with all the dangers which that entailed), Cook made the decision to turn west. Between seven and eight o'clock that evening, he had reached the latitude of 60 degrees and ten minutes – placing the vessel around 750 kilometres from Antarctica.[48] Yet, despite venturing this far south, there was no sign of the Great Southern Continent. There were no currents to suggest a nearby landmass and Cook ordered a line to be lowered to see the depth of the seabed but, 240 metres down, it still did not reach the

46. Cook's Journal, 25 January 1769.
47. Cook's Journal, 25 January 1769.
48. Cook's Journal, 30 and 31 January 1769.

sea floor, again indicating that any sizeable landmass was unlikely to be in this vicinity.

For the next two months, the *Endeavour* sailed northwest into the warmer regions of the South Pacific Ocean, arriving at Vahitahi (part of the Tuamotu archipelago) on 4 April, and, for the next week, journeyed past several smaller islands before reaching Tahiti. This was the destination where Cook was able to fulfil the first set of instructions: the observation of the transit of Venus. The *Endeavour*'s crew constructed an observation post on a peninsula which they named Point Venus and, in their spare moments, variously quibbled and fraternised with the locals. Cook used his time more productively, producing a detailed set of maps of several islands in the vicinity, with the assistance of Tupaia – an expert indigenous navigator from the nearby island of Ra'iātea. Tupaia's training had been undertaken in the traditional school of learning, in which navigation (mainly by stars) was committed to memory and was inseparable from other areas of study, such as religion, genealogies, proverbs and histories.[49] The meeting of these two men represented the convergence of distinct navigational traditions, not to mention the more immediate cultural differences that had the potential to confound their cooperation. Cook regarded Tupaia as 'a very intelligent person' and 'the likeliest person to answer our Purpose'[50] – that purpose being expertise about the geography and peoples of the region. However, the *Endeavour*'s captain initially displayed some reticence about having Tupaia accompany the expedition on its next leg, though, and it was left to Banks to persuade Cook of the advantages that this local expert would be to the voyage, along with Tupaia's own earnest pleas to be part of the journey.

However, while Banks was a forceful advocate for Tupaia's involvement in the expedition, he in no way considered the Tahitian navigator as his equal. Yes, Banks wrote in his journal that Tupaia was 'certainly a most proper man, well born, chief ... or preist of this Island' and that he was:

> skilld in the mysteries of their religion; but what makes him more than any thing else desireable is his experience in the navigation of these people and knowledge of the Islands in these seas; he has told us the names of above 70, the most of which he has himself been at.

49. A. Salmond, *Aphrodite's Island: The European Discovery of Tahiti* (Berkeley: University of California Press, 2009), pp. 33 ff.
50. Cook's Journal, 13 July 1769.

However, once Tupaia joined the expedition, Banks regarded him with extreme condescension, almost as if he was some exotic chattel. 'I do not know why I may not keep him as a curiosity', the botanist proposed,

> as well as some of my neighbours do lions and tygers at a larger expence than he will probably ever put me to; the amusement I shall have in his future conversation and the benefit he will be of to this ship, as well as what he may be if another should be sent into these seas, will I think fully repay me.[51]

By mid-August, the *Endeavour* was once more heading due south. Cook had by now read his second set of instructions and was venturing in search of the Great Southern Continent. On 1 September, the ship had reached 40 degrees latitude, 'having not the least Visible signs of land'. There is a sense that Cook had doubts about the existence of this supposed continent, as he had expressed shortly after leaving Rio de Janeiro. The following day, a violent storm moved rapidly into the area, giving Cook the pretext that he needed to alter course. As the tempestuous weather engulfed the ship, he turned north, 'least we should receive such Damage in our Sails and Rigging as might hinder the further Prosecutions of the Voyage',[52] and then for the next month, headed west. By the end of September, the Endeavour was just above the fortieth parallel, making its way westward towards New Zealand. Cook knew the rough coordinates of the territory's west coast from Tasman's map, but had no idea where its east coast lay (which was the direction he was sailing towards). All previous cartographers had been content to leave the shape of New Zealand as a single length of coastline which Tasman had encountered from the west. The form of the country was rightly left incomplete (with the exception of Bellin's 1753 *Carte réduite des terres Australes*, which speculated that New Zealand was part of the Great Southern Continent). However, as Cook neared the vicinity of the country, there was nothing in his journal that suggests he anticipated such a vast landmass ahead.

In the first few days of October 1769, some of the oldest navigational techniques alerted the *Endeavour*'s crew that land was ahead. Banks spotted a piece driftwood bobbing around in the water, unknown species of birds began to appear and a storm that hit the ship with great violence, and that moved on in a few minutes, was another indication

51. Bank's Journal, 1 July 1769.
52. Cook's Journal, 2 August 1769.

known to seasoned sailors that land was close. '[S]uch squalls', Banks noted, 'are rarely (if ever) met with at any considerable distance from it [land].[53] On 6 October, Cook wrote that the sea appeared 'paler than common, and hath been so for some days past'.[54] All the omens pointed to the *Endeavour* getting close to a significant body of land ahead.

53. Bank's Journal, 3 October 1769.
54. Cook's Journal, 6 October 1769.

Chapter 8

Mapping the East Coast of New Zealand

Cosmography and Mapmaking Methods

Eighteenth-century cartographers – Cook included – saw themselves very much as disciples of science and, more specifically, the particular branch known as cosmography (a term popularised in English in the 1650s by the amateur geographer Peter Heylyn, who incidentally was one of the advocates of the existence of the Great Southern Continent).[1] This was an approach to understanding the physical world that incorporated geography, astronomy, mathematics, geometry and cartography into a sort of hybrid discipline of its own,[2] rather than having them fenced off as separate areas of study. A century after Heylyn's work appeared, the German mathematician and astronomer Johann Doppelmayr and the French geographer Philippe Buache independently published works that reached very similar conclusions on the importance of cosmography in synthesising the discipline of mapmaking with other areas of knowledge.[3]

Cartographers in this era almost instinctively drew on an extensive palette of the sciences and the arts when producing their maps. It was up to the individual mapmaker to choose which configuration of various disciplines they used to produce their works and, in this process, they were guided by that compelling Enlightenment-era impulse to

1. P. Heylyn, *Cosmographie, in Four Books. Containing the Chorographie and Historie of the Whole World*, 2nd edn (London: Henry Seile, 1657), p. 1092.
2. M. Edney, 'Mathematical Cosmography and the Social Ideology of British Cartography, 1780-1820', *Imago Mundi* 46, no. 1 (1994), p. 101.
3. Ibid., p. 102.

accumulate and categorise information. Mapmakers disclosed locations and landforms that were previously unknown to Europe, while botanists, artists and others unveiled the attributes of those territories, and nearly all of them typically produced text to accompany their work.

Yet, for all the pretensions to scientific purity, there was also a theological dimension to such undertakings, even if it appeared submerged by the day-to-day work of cartographers. In discovering and reporting on new places, what was also being revealed were layers of Creation. This was known at the time as physico-theology – a belief that the world and everything in it was not only part of God's purpose, but also verification of it.[4] In this view of Creation, the knowledge of nature and the knowledge of God walked hand in hand. As John Ray, a Fellow of the Royal Society, suggested in 1717, explorers and scientists ought to be 'ingenious and industrious persons', who 'delighted in searching out these natural Rarities and observing the outward Form, Growth, Natures, and Uses' of the natural world, for the ultimate purpose of 'reflecting upon the Creator of them his due Praises and Benedictions'.[5] This conception of nature was summed up by the eminent Swedish botanist Carl Linnaeus, who famously concluded that '[t]he Earth's *Creation is* the glory of God, as seen from the works of Nature',[6] and so studying the world and classifying its constituent parts would help to reveal its divine order. Daniel Solander – one of Linnaeus' greatest protégés, and a colleague of Banks – was deeply imbued with this approach to information-gathering and took it with him as a piece of ideological luggage during the *Endeavour*'s expedition. In this way, cosmography had sacred as well as secular functions and, to that extent, Cook's relentless mapping throughout the *Endeavour*'s voyage was contributing to religious as well as scientific and imperial ambitions. The boundaries separating these three motives were barely visible in practice and frequently surmounted. So, while, theologically, maps acted as a form of confirmation in the divinity of Creation, strategically, the charting of each new territory forced European powers to reconsider the nature of their global interests and the possible threats to those interests. Commercially, extensions of

4. W. Derham, *Physico-Theology: Or, A Demonstration of the Being and Attributes of God* (London: W. Innys, 1713).
5. J. Ray, *The Wisdom of God Manifested in the Works of the Creation* (London: R. Harbin, 1717), p. 218.
6. 'Finis creationis telluris est gloria Dei ex opere naturae per hominem solum', in C. Linnaeus, *Systema naturae per regna tria naturae* (Stockholm: Laurentii Salvii, 1758), p. 3.

areas in the world that were mapped potentially represented new trading routes and new markets, and, for humanitarian and missionary groups, these unfamiliar places outlined on maps appeared as opportunities for civilising and converting the inhabitants that might be living there.

Of course, the mapping of New Zealand – a place that seemed geographically remote, strategically unimportant and commercially unrewarding – might have appeared to bypass some of the elements of the relationships of power between Europe and the rest of the world. There was, for example, no effort to impose arbitrary borders in New Zealand in the manner that cartographers were beginning to do in Africa. The territory that Tasman had first partially outlined was not about to be carved up for consumption by rival imperial powers, let alone populated by them. Yet, the fact that New Zealand's aquatic boundaries were in the process of being charted was just as much the imposition of a border as those lines that were arbitrarily compartmentalising other parts of the world. Outlines implied boundaries, and even statehood. Moreover, all the time, it was what Europe knew that counted when maps were being made. The blank spaces in the Southern Hemisphere, like those in much of Africa in this period, were labelled with the word 'incognita' – unknown.[7] It did not matter that there might be people already living in such locations who certainly knew of them. It was Europe which judged what constituted 'being known'.

It would be an exaggeration to imply that all mapping was part of a preordained process that would end in direct colonial authority. Tasman's map of New Zealand is a case in point. The Dutch East India Company went to considerable expense for the map to be produced, yet it never made any attempts at exercising imperial influence over the territory whatsoever. There was no formula by which cartography inevitably led to conquest. However, it is just as important not to discount the effects of mapping on what can be called the 'imperial imagination'. European officials, politicians and that expanding sea of middle-class readers saw areas marked 'incognita' as a mystery and wanted them instead to be 'cognita'. Furthermore, while pragmatists might be deterred from intervention by the vast distances separating Europe from some unclaimed territories, and the colossal expense that would be involved in asserting a form of dominion over them, for many others, the imperial imagination nudged people towards thoughts of intervention.

---

7. J.C. Stone, 'Imperialism, Colonialism and Cartography', *Transactions of the Institute of British Geographers* 13, no. 1 (1988), p. 58.

In the context of the imperial imagination, both the timing and extent of Cook's expeditions have great significance. When Cook departed Plymouth in the summer of 1768, a third of the Earth's surface was still practically unknown to Europe.[8] His voyages into parts of these uncharted and mysterious regions coincided with recent advances in navigational technology, which lent much greater certainty to the locations that were appearing on maps. (Cook was one of the few navigators on earth who had mastered the lunar distances as a means of calculating longitude.[9]) The result was that the hidden face of much of the South Pacific Ocean (including the absence of the putative Southern Continent) was revealed and the mystery shed.

Cook contributed nothing to the technology of mapping or navigation, but his expertise in applying the latest scientific methods and instruments to cartography, along with his fastidious commitment to accuracy, became hallmarks of his mapmaking.[10] (Some of his charts were so precise that they were not resurveyed until 1996.[11]) The presence of Charles Green on the first voyage to New Zealand ought not to be eclipsed, though, by Cook's luminescence. Green had been appointed as an assistant to the Astronomer Royal in 1761 and played a vital role in checking and confirming Cook's coordinates, which Cook openly acknowledged:

> In justice to Mr. Green, I must say that he was indefatigable in making and calculating these observations, which otherwise must have taken up a great deal of my time, which I could not at all times very well spare; not only this, but by his instructions several of the petty Officers can make and calculate these observations almost as well as himself. It is only by such Means that this method of finding the Longitude at Sea can be put into universal practice.[12]

---

8. B. Greenhill, 'Captain James Cook, RN', address at Cook Commemorative Service, Westminster Abbey, 11 February 1979, p. 1.
9. G.W. Littlehales, 'The Decline of the Lunar Distance for the Determination of the Time and Longitude at Sea', *Bulletin of the American Geographical Society* 41, no. 2 (1909), pp. 84-85.
10. E.G.R. Taylor, 'Navigation in the Days of Captain Cook', *The Journal of Navigation* 21, no. 3 (1968), p. 256.
11. J. Lockett, *Captain James Cook in Atlantic Canada* (Halifax: Formac Publishing, 2010), p. 7.
12. Cook's Journal, 23 August 1770.

As the *Endeavour* approached the eastern shores of New Zealand, the scientific, navigational, cartographical and philosophical planets were in perfect alignment. At no previous time had the conditions for a successful mapping of the territory been so well-suited to the task ahead. Cook's journal entry as New Zealand was sighted by a European for the first time in 127 years was a study in discipline. There was no elation, let alone an affected sense of destiny. Instead, Cook made the following typically clinical comment:

> At Midnight brought too and sounded, but had no ground with 170 fathoms. At daylight made sail in for the Land, at Noon it bore from South-West to North-West by North, distant 8 Leagues. Latitude observed 38 degrees 57 minutes South; Wind North-East, South-East, Variable; course South 70 degrees West; distance 41 miles; latitude 38 degrees 57 minutes observed South; longitude 177 degrees 54 minutes West.[13]

Cook's Approach to Charting

Even before New Zealand was first spotted by the *Endeavour*'s crew, Cook knew that it was unlikely to be part of some immense undiscovered landmass. As he set his course to due west from 28 September (on the same latitude as what he would soon name as Poverty Bay), every day without reaching the territory diminished its potential size. Tasman's map identified the location's western extremities, and Cook was now reaching it from its eastern side. Having fulfilled that part of the secret instructions which required him to search for the Southern Continent (and being reasonably certain that it did not exist in the way the Admiralty had imagined), Cook was now expected to trace thoroughly the outline of New Zealand (filling in the blanks that Tasman had left) for as long as provisions on the ship would support this undertaking.[14]

On 9 October, Cook sailed towards a large bay but, as night descended, the direction of the wind made getting any nearer too risky. The next morning, Cook tacked closer and made some brief observations about the contours of the land before starting the business of mapping the country. Being anchored off what is now Gisborne, he began taking

13. Cook's Journal, 7 October 1769.

14. Commissioners for executing the office of Lord High Admiral of Great Britain, 'Secret. Additional Instructions'.

Official portrait of Captain James Cook (Nathaniel Dance-Holland, 1775)
National Maritime Museum, United Kingdom

measurements of the depth of the bay ('10 fathoms [with] a fine sandy bottom ... distance from the Shore half a League').[15] Some of the crew, including Cook and Banks, then went on shore, and had their first encounter with Māori. It was one of those typically clumsy interactions between two peoples whose cultures were utterly alien to each other. For the visiting party, although curiosity was tempered with caution, it was not long before a stray spark of misunderstanding ignited a short but lethal conflict. What the *Endeavour*'s crew regarded as an exploratory landing on the coast was interpreted by the Māori residents as a potential invasion.[16] The ensuing clash between the two groups resulted in one Māori being killed. The following day, Cook again went

15. Cook's Journal, 9 October 1769.
16. Banks' Journal, 6 November 1769; G. Obeyesekere, '"British Cannibals": Contemplation of an Event in the Death and Resurrection of James Cook, Explorer', *Critical Inquiry* 18, no. 4 (1992), 645-46.

on shore, this time with Tupaia as translator and a body of marines for protection. After some exchanges, and the performance of a *haka* (war dance) by members of this *hapū* (subtribe), it looked as though peace had triumphed. However, it was again short-lived. More misunderstandings led to more Māori being killed by Cook's men.

From the perspective of the local *hapū*, the *Endeavour* was seen as a great bird, with its landing vessel perceived as a fledgling with 'a number of partly-coloured beings, but apparently in the human shape, also descending', as the trader Joel Polack, who interviewed some of the descendants of this encounter, wrote in the 1830s. Māori consternation soon gave way to apprehension, though:

> [T]he bird was regarded as a houseful of divinities. Nothing could exceed the astonishment of the people. … Many of these natives observed, that they felt themselves taken ill by only being particularly looked upon by these Atuas [deities]. It was therefore agreed, that, as these new comers could bewitch with a single look, the sooner their society was dismissed, the better it would be for the general welfare.[17]

Yet, following this deadly exchange, and the ensuing heightened mutual suspicions, there was a reconciliation of sorts, with Cook attempting to patch up the relationship – which achieved at least a partial success.[18]

The subsequent meetings between the *Endeavour*'s crew and Māori – most of which were mediated by Tupaia – have been exhaustively documented.[19] Cook, however, remained focussed on the immediate task at hand: to map the country. This would be a collective effort and, in keeping with the nature of cosmography, would draw on the skills of the various experts on the ship to assemble as much information about the territory as possible. The edifice of intelligence that was built up as the *Endeavour* circumnavigated New Zealand for the next six months was founded on Cook's maps, but would be augmented with anthropological, botanical, geographical and geological information,

17. J. Polack, *New Zealand: Being a Narrative of Travels and Adventures during a Residence in that Country*, 2 vols (London: Richard Bentley, 1838), Vol. 1, pp. 15-16.
18. Cook's Journal, 10, 11 October 1769.
19. Perhaps most exhaustingly in A. Salmond, *Two Worlds: First Meetings between Maori and Europeans, 1642-1776* (Auckland: Penguin, 2018), pp. 119 ff.

which would give the Admiralty a much more thorough impression of the territory.

* * *

Cook's approach to bringing into being a complete map of New Zealand started with charts he prepared for specific parts of the territory's coast. Twenty-seven of these have survived from this expedition and fall into one of three general categories. The first – three of them – are untitled draft works. The next five are 'intermediate charts' which depict the outline of lengths of the coast and which were redrawn in a reduced scale. The final category are the completed charts, marked out in pen and then tinted in portions with diluted ink – a process known as pen and wash. These are the maps containing the numerous place-names that Cook affixed to locations around New Zealand, along with seabed depths, the course that the *Endeavour* followed and various lines of latitude and longitude.[20]

The importance of the three draft charts is that they illustrate – literally – how Cook went about producing his maps. They are, in that sense, the raw material of all his later cartographical accomplishments. It is not possible to recreate the order of every single pen stroke that Cook made, nor the sequence of measurements he put to paper, the draft calculations he threw away when he had finished with them, or the sketches he may have begun but then disposed of. However, there are some aspects of his mapmaking method that can be reconstructed.

Firstly, once the *Endeavour* was anchored, usually a few kilometres offshore, the ship's location would be established. Then, Cook would instruct his officers to select particular points along the coast and, using his azimuth compass, he would get his ruler and draw with a pencil on his draft chart a series of intersecting lines based on the position of the ship, the angle of some celestial body (often the sun) to the horizon, and the distance from the coast.[21] These draft maps were made on a scale of one inch to three nautical miles. Cook would then reduce this in his intermediate charts to a scale of one inch to five nautical miles.

20. J.R.H. Spencer, 'A New Zealand Draught Chart by James Cook, RN', *Cartography* 13, no. 4 (1984), p. 316.
21. D.J. Warner, 'True North: And Why It Mattered in Eighteenth-Century America', *Proceedings of the American Philosophical Society* 149, no. 3 (2005), p. 374; Spencer, 'A New Zealand Draught Chart by James Cook, RN', p. 316.

Then, as the *Endeavour* sailed along New Zealand's shore, Cook would use a position he fixed as the starting point for measuring stretches of the coast. This would involve the process known as dead reckoning, which required information on the vessel's speed, direction and the distance it had travelled over a given time.[22] (Cook relied on 30-second intervals, for which he used an hour-glass calibrated for this function.) A knotted line attached to a plank of wood or log (for the purpose of acting as a sea anchor) was put in the water and unwound from the stern of the ship in order to measure the distance travelled over a set time, while accommodating factors such as currents and winds, which tended to be more pronounced the closer a vessel was to the shore. Also, in addition to relying on measurements from the knotted line, Cook had to make corrections to take into account leeway (which is how far the vessel had deviated from the course being steered[23] – sometimes referred to as 'lateral drift').[24]

From combining the screeds of coordinates and other measurements with estimates of the ship's drift, scrutiny of features of the coastline, low-water-mark soundings[25] and periodic visits on shore Cook was able to compile his completed charts of the country. For the accuracy of his maps, he depended on the frequency and precision of the data he collected. However, this was far from being a straightforward process of converting calculations into lines on a map, with less easily quantifiable skills, such as intuition and judgement borne from experience, contributing to the process.

There was then a series of stages Cook followed to produce the final version of his charts. The first of these was to sketch the outline of the coast and provide a depiction of the topography that was immediately visible, using hachure lines (short pencil lines with which he sketched the relief of the terrain that bordered the coast). These did not have the precise measurements of contour lines but that was not their purpose. Instead, they were meant to be indicative of the general landforms in the area. The fullness of this topographical detail, enhanced with the use of green or brown brushwork (in the style of military maps in this

22. E.G.R. Taylor, 'Five Centuries of Dead Reckoning', *The Journal of Navigation* 3, no. 3 (1950), 280-85.
23. W. Glover, 'The Eighteenth Century Practice of Navigation as Recorded in the Logs of Hudson's Bay Company Ships', *The Northern Mariner/Le marin du nord* 26, no. 2 (2016), p. 159.
24. Spencer, 'A New Zealand Draught Chart by James Cook, RN', p. 319.
25. T.M. Knight, 'Cook the Cartographer', *Cartography* 7, no. 3 (1971), p. 113.

period), were among those features that made Cook's maps distinctive for this time.[26]

Then, once he was satisfied with what he had sketched, he went over the coastal outline in ink and, as a final touch, he used various tints of ink-wash to accentuate the relief that he had drawn in hachure strokes.[27] The result was a set of works that were unequalled in their detail and accuracy, and that also held some aesthetic appeal. The sinuous coasts that Cook drew in thick black ink lines were the boundaries of a land whose interior form still remained mysterious, but for which there were now hints of its relief. Undulating hill country, vacant plains, mountain ranges and rivers braiding out to sea were all captured by Cook. His clinical reduction of three-dimensional geography to paper – the essence of all mapmaking – was juxtaposed with the slightly ethereal expanses of blank space which it was left to the imagination to fill until such time as later explorers travelled inland to survey them.

Cook's decision to apply colour to his maps (albeit in anaemic watery tones) was unusual for captains or onboard cartographers in this era. The normal expectation was that the cartographer would produce their maps in black ink, after which the engraver and publisher would get to work applying colours, often as much to embellish aesthetically the image as to enhance the topography being depicted. However, now, Cook was assuming the role of chromatic commander, producing pigmented maps that captured what he saw first-hand as he charted New Zealand's coast, rather than leaving such decisions to printers back in Britain.

Inevitably, there were occasional imperfections. Cook mapped Banks Peninsula as an island and, conversely, Stewart Island as a peninsula. However, in both cases, these errors can be attributed to the constraints of time that did not allow him to explore every length of coast ahead of him. Yet, even in this, Cook can be partly excused because given the mapping that he did conduct in those locations, both errors were, in fact, reasonable deductions for him to make in the circumstances. The luxury of exploring every bay, inlet and crevice of the coast was not available to him. This was an expedition of perpetual motion, with the captain having to balance a range of considerations – from weather, food supplies and the morale of his men, to the overall length of his voyage, the requirements to carry out his instructions and the differing expectations of the various scientists on board. And, despite the very occasional missteps in his mapping, Cook took on the role of charting

26. Ibid.
27. Spencer, 'A New Zealand Draught Chart by James Cook, RN', p. 319.

the territory's coast with unrelenting precision, which was his primary objective. He was fully seized of the need to be 'indefatigable in making and calculating these observations' and relied on some of the officers on board to assist in this work, 'which otherwise must have taken up a great deal of my time, which I could not at all times very well spare'.[28] The task that lay before Cook was substantial: to apply all his cartographical and navigational skills, along with those of the officers on the *Endeavour*, to bring form to New Zealand from the void of the unknown.

Life Onboard the *Endeavour*

It would take no great act of divination to predict that nearly 100 men confined to a small ship for what felt like endless weeks on the ocean, with ample supplies of wine, beer and rum, would be an intoxicating mix. Drunkenness on lengthy sea voyages had been a feature of such journeys for centuries, but alcohol continued to be regarded as an essential supply because it was still the most effective antidote to the dreariness of long-distance expeditions. The excess consumption of alcohol had become one of those necessary vices that captains had to contend with but could never quite purge. In one of the more serious episodes of inebriation, as the *Endeavour* made its way towards New Zealand, Banks dispassionately recorded that 'the seamen Rayden ... was this morn found so drunk that he had scarce any signs of life and in about an hour he expird'. Banks went on to note, following this fatality, that 'there is not a cask on board the ship that has not been tap'd to the great dissatisfaction of the owners'.[29] Yet, even death was not sufficient to have a sobering effect on the crew. That Christmas, according to Cook, just about everyone on board got drunk. Moreover, later in the voyage, when dozens of the *Endeavour*'s crew succumbed to some sort of illness, the captain laconically observed that 'every one [is] sick except the Sailmaker, an old Man about 70 or 80 years of age; and what is still more extraordinary in this man is his being generally more or less drunk every day'.[30] Cook may have disapproved of the way alcohol was abused on board but, in general, tolerating it was the lesser of two evils. Prohibition certainly would have been inconceivable on ships in this era.

Drunk and unruly sailors were one of the aspects of ship life that impinged on Cook's work and that added to the pressure he was already

28. Cook's Journal, 27 August 1770.
29. Banks' Journal, 28 August 1769.
30. Cook's Journal, 26 January 1771.

under to complete his charts of New Zealand within the limited time allotted to his circumnavigation of the territory. Illness was another. Just a fortnight before reaching the territory, Banks scrawled in his journal that he was 'still very sick at the stomach which continualy supplys a thin acid liquor which I discharge by vomit'.[31] A week later, he had recovered enough to attend to another ill member on board – Solander – who had been unwell for several days from the effects of scurvy.[32] Later in the expedition, Solander was afflicted with another sickness. He was 'irregularly intermitting fever and violent pains in his bowels' and appeared 'very much emaciated by his tedious Illness'.[33] Inevitably in this era, lengthy voyages were accompanied by deaths. Two of the more prominent members of the *Endeavour* expedition, Green and Tupaia, succumbed to disease and did not live to see the end of the voyage, and the ship's hydrographer, Robert Molyneux, along with one of the officers, Zachery Hicks, also passed away before the vessel returned to England, while others died from drowning and mishaps on board.

As he sailed from England in 1768, Cook would have anticipated that not all those on the ship would be returning with him and this awareness removed the possibility of carrying out his instructions at a more leisurely pace. As a rough rule, the longer a vessel was at sea, the higher the rate of fatalities would be so, amidst all the daily measurements and calculations in which Cook was immersed, he also had to keep an eye on the overall length of the expedition, and somehow estimate how long was a suitable time to remain at each location. There was no fixed rule for this and, like so many other aspects of this venture, it relied in Cook's well-tuned sense of judgement.

Above all, mapping New Zealand was very much a joint venture. For all of Cook's towering cartographical talents, there were others on the *Endeavour* whose contribution was vital to the comprehensive impression of the territory that was slowly being assembled. The artist Sydney Parkinson, for example, drafted detailed sketches of the appearance of the coast from various points where the ship anchored.[34] He also recorded the sort of information that had been requested in the Admiralty's instructions and fleshed out the impression of New Zealand that was provided on maps. One typical example of this was during the

31. Banks' Journal, 16 September 1769.
32. Banks' Journal, 23 September 1769.
33. Ibid., 31 March 1771.
34. S. Parkinson, *A Journal of a Voyage to the South Seas, in His Majesty's Ship, the Endeavour* (London: Stanfield Parkinson, 1773), p. 86.

Endeavour's stay at Poverty Bay, where Parkinson went on shore and jotted down his observations of the surrounds:

> We found here a sort of long-pepper, which tasted very much like mace; a Fulica, or bald Coot, Of a dark blue colour; and a Black-bird, the flesh of which was of an orange colour, and tasted like stewed shell fish. A vast quantity of pumice stone lies all along upon the shore, within the bay, which indicates that there is a volcano in this island.[35]

While Parkinson was harvesting details on the physical appearance of the territory, Green was serving a more direct role in tracing New Zealand's outline. Along with Cook, Green was heavily involved in taking measurements of longitude. Cook frequently referred to the astronomer's work, indicating how much time Green saved him by assisting with the seemingly endless stream of calculations that were required to establish the ship's position. A typical example from Cook's journal reveals the extent of Green's contribution to this task:

> The greatest differance between any two – viz., the first and third – is but 26 minutes, and the mean of these two differ from the mean of the whole only 2 minutes 26 seconds. This shews to what degree of accuracy these observations can be made even by Different Persons, for four of these were made and computed by Mr. Green and the rest by myself. The Longitude given by the Ship, reckoning from the last Observation 5 Days ago, differs only 8 Miles from the Observation.[36]

Others accompanying the expedition were also proving their worth. The hydrographer Richard Pickersgill, who was nineteen when he joined the voyage, became a reliable assistant to Cook, as did Molyneux, whom Cook described as 'a young man of good parts', who had 'unfortunately given himself up to Extravagancy and intemperance, which brought on disorders that put a Period to his Life'.[37]

Despite his youth, Pickersgill was an accomplished cartographer in his own right. As the *Endeavour* sailed south after making landfall at Poverty Bay, it was Pickersgill who made a chart of this portion of the

35. Ibid., p. 89.
36. Cook's Journal, 29 December 1768.
37. Cook's Journal, 16 April 1771.

coast. It was the first few days of exploring and mapping of the territory, and there was still uncertainty as to the form of New Zealand and, particularly, whether it was part of a much greater landmass. Pickersgill believed that it was and, accordingly, entitled his map, *A chart of part of the So. Contit. between Poverty Bay and the Court of Aldermen*. It contained a careful outline of the coast and some longitude markings, as well as a comment added months later (possibly by Pickersgill, although it cannot be known for certain), which reads: 'This chart was taken before this country was found to be an Island.'[38] This was one of twelve charts that Pickersgill produced during the expedition, many of which were sources of reference when the final printed versions of maps of New Zealand were being assembled back in Britain. He also made his own maps based on Cook's charts, although the purpose for doing so is unclear.[39]

The naturalist and instrument maker Herman Spöring, who was part of Banks' entourage, also contributed – on a less significant scale – to the corpus of sketches, measurements and charts that would be used to form New Zealand's first complete map. He drew a series of coastal views of Poverty Bay, which Banks then labelled, describing the various geographical features on view (although when it came to engraving and publishing these images, Banks ignored Spöring's sketches in favour of those produced by Parkinson).[40] Ironically, while Banks favoured what he felt was Parkinson's more aesthetic approach to depicting this part of

38. R. Pickersgill, *A chart of part of the So. Contit. between Poverty Bay and the Court of Aldermen discovered by His Maj.s Bark Endeavour* [copy of ms map] [1769], National Library of New Zealand, ref. MapColl-832.1aj/[1769]/Acc.12471.
39. R. Pickersgill, *New Zealand: Capt. Cook's Discoveries* [1771], National Library of Australia, Charts of the Hydrographic Department (as filmed by the AJCP) [microform] : pre-1825 :[M406], 1770-1824./File 139-40.458; J.R.H. Spencer, 'The New Zealand Charts Compiled on HM Bark Endeavour by Cook, Molyneux and Pickersgill, 1769-70', *Archifacts: Bulletin of the Archives and Records Association of New Zealand*, no. 1 (March 1985), pp. 2-3; B. Keir, 'Captain Cook's Longitude Determinations and the Transit of Mercury: Common Assumptions Questioned', *Journal of the Royal Society of New Zealand* 40, no. 2 (2010), p. 34.
40. S. Parkinson, *A View of the North Side of the Entrance into Poverty Bay & Morai Island in New Zealand. 1. Young Nick's Head. 2. Morai Island. View of another Side of the Entrance into the said Bay*, S. Parkinson del. R.B. Godfrey sc. Plate XIV [London, 1784], National Library of New Zealand, ref. PUBL-0037-14.

New Zealand's coast, Spöring in effect had the last laugh; his much more artistic *A Fortified Town or Village called a Hippah*[41] achieved lasting popularity after it appeared in John Hawkesworth's subsequent multi-volume work detailing Cook's expedition[42] and enjoyed innumerable reprints ever since.

What all this sketching, painting, outlining, drawing, measuring and calculating goes to show is that the production of the first full map of New Zealand was a collective enterprise. Cook rightly stood on the pinnacle of this accomplishment but beneath him were men not only with the range of skills, but also with a variety of outlooks on the natural world, extending from the stiff mathematical formality of Cook's charts, all the way through to Spöring's more supple renditions of the landscape. The particular combination of skills of those on board the *Endeavour* was brought together as much by circumstance as design but, under Cook's direction, he would harness all that was available to him in pursuit of what would become one of the greatest mapping exercises in history.

---

41. H. Spöring, *A Fortified Town or Village called a Hippah, built on a Perforated Rock, at Tolaga in New Zealand*, Morris sculp. (London. Alexr. Hogg [1784?]), National Library of New Zealand, ref. B-098-025.
42. Aspects of the gestation of Hawkesworth's books are dealt with in J.L. Abbott, 'John Hawkesworth: Friend of Samuel Johnson and Editor of Captain Cook's Voyages and of *The Gentleman's Magazine*', *Eighteenth-Century Studies* 3, no. 3 (1970), pp. 341-42.

Chapter 9

One or Two Islands Separated by a Strait?

Cape Turnagain

While the *Endeavour* was anchored at Poverty Bay, the crew could have been forgiven for thinking that this was another instance of business as usual on their South Pacific odyssey. Under Cook's direction, they were following the established pattern of sighting land, getting bearings, cautiously setting foot on what looked to be the nearest safe stretch of shore, gathering plant specimens, either clashing or cooperating with the local indigenous community and, generally, garnering as much information as possible in whatever time they had allotted at that location. When the captain decided, they would sail to somewhere nearby and repeat it all over again. For Cook, however, New Zealand presented itself as a cartographical challenge, principally because he had 'no time to spare',[1] as he later put it. So, in his mind, he was continuously trying to balance the need for maintaining a tight schedule with the seemingly opposing requirement to assemble a comprehensive portrait of the territory that the Admiralty expected of him. This was a balancing act that he alone was responsible for managing.

There was an added layer of complexity for Cook to contend with, though, and this appeared in the form of the puzzle left by Tasman. As Cook studied the 137-year-old chart of New Zealand that had been produced by his Dutch predecessor, he was forced to make some educated presumptions that would influence the direction of his coastal exploration. The more he examined Tasman's map, the more he was able

1. Cook's Journal, 11 August 1770.

to distil the puzzling aspects of it into a few very clear questions and suppositions, all of which related to the shape of the territory.

Gazing at that part of New Zealand's western littoral that Tasman had drawn, the first presumption Cook made was that what the Dutch captain had marked as the northernmost point of the territory – Cape Maria van Diemen – was almost certainly the termination point of the country in that location. Cook expressed no thoughts about New Zealand extending further north beyond that position. This left two large questions: what was the form of the southern reaches of the territory? (Did it, for example, widen and connect to the Great Southern Continent?); and what was the nature of the waterway north of Murderers' Bay? Was it some vast gulf, or was it part of a passage that separated New Zealand into two territorial bodies? Given sufficient time, Cook could answer these questions with complete certainty. However, having been deprived of that luxury, he was forced to make a series of decisions that he hoped would accelerate the process of mapping the country without having to sacrifice accuracy to haste.

This was the nature of distant territorial cartography at this time. It was not one of those scientific undertakings that could be conducted at leisure in a laboratory or library by gentlemen scholars. Instead, this sort of work was at the mercy of the elements, hostile peoples, sickness and that specific form of apprehension that always overshadows efforts to probe the unknown. As soon as Cook anchored at Poverty Bay, he was confronted with a dilemma: should he sail north or south? If he chose the former, he knew, based on Tasman's map, that it was roughly an 800-kilometre voyage to reach Cape Maria van Diemen. This was certainly the most predictable of the two options to follow. If he decided to sail south, however – especially if it turned out that New Zealand was a vast peninsula jutting out from the Great Southern Continent – then he risked heading into the frozen regions of the Southern Ocean, tracing an interminable coast in highly unfavourable conditions and all the while heading further away from some of the other locations that were the object of this expedition to explore. After nearly three days in Poverty Bay, Cook exercised his Solomonic wisdom and resolved on his next course: 'My intention is to follow the direction of the Coast to the Southward, as far as the Latitude of 40 or 41 degrees, and then to return to the Northward, in case we meet with nothing to incourage us to proceed farther.'[2]

What was it about this particular latitude that made him act so resolutely? Again, it goes back to Tasman's map. As he examined it, Cook

2. Ibid., 11 October 1769.

could see that the area of 40 to 41 degrees was approximately the latitude of Murderers' Bay and Cape Pieter Borrel. Tasman's chart implied that these two points were at the opening of an enormous bay (the mouth of which was about 150 kilometres wide). However, Cook's instinct told him that it was a potential strait and, in order to discover if this was the case, the logical thing to do was to sail to the same latitude on the east coast, which would reveal whether there was an opening or not that joined the corresponding area on the west coast that Tasman had mapped.

For the next several days, the *Endeavour* crept along the coast, even sailing at night when it was possible in order to make the best use of the available time. However, when the ship reached the latitude where Cook predicted an opening to the west would be, there was none, and his stoic response only just concealed his frustration:

> Seeing no likelyhood of meeting with a Harbour, and the face of the Country Visibly altering for the worse, I thought that the standing farther to the South would not be attended with any Valuable discovery, but would be loosing of Time, which might be better employ'd and with a greater Probability of success in examining the Coast to the Northward. With this View, at 1 p.m. Tack'd and stood to the Northward. ... At this time we could see the land extending South-West by South, at least 10 or 12 Leagues. The Bluff head or high point of land we were abreast off at Noon I have called Cape Turnagain because here we returned. It lies in the Latitude of 40 degrees 34 minutes South. ... The ridge of Mountains before mentioned extends to the Southward farther than we could see, and are every where Checquer'd with Snow.[3]

Cook had kept coy about his belief that there was a strait that led to Murderers' Bay.[4] He sensed that it existed yet, when he had sailed to the area where it ought to have appeared, all he was confronted with was an impenetrable mountain range deep inland, sweeping southward. Maybe it was a hint that this was indeed part of the Great Southern Continent and that the mountainous terrain that extended as far as the eye could see was part of its extremity. Although Cook had his doubts about the existence of such an enormous landmass, privately, he was much more frustrated with his judgement that for once had let him down. Three

3. Ibid., 17 October 1769.
4. Hooker, 'James Cook's Secret Search in 1769', p. 301.

months later, when he finally resolved the question of whether there was a strait connecting the east to the west around this latitude, Banks recorded how Cook was unable to conceal his delight. His instincts had been correct all along:

> the captn went to the top of a hill and in about an hour returnd in high spirits, having seen the Eastern sea and satisfied himself of the existence of a streight communicating with it, the Idea of which had Occurd to us all from Tasmans as well as our own observations.[5]

By the end of October 1769, Cook had anchored at Tologa Bay, a sailing distance of around 50 kilometres north of Poverty Bay. His journal entry for this location was typical of the mix of content he included that ultimately would be aggregated to develop his map of New Zealand. He wrote that Tolaga Bay:

> is moderately large, and hath in it from 13 to 8 and 7 fathoms, clean sandy bottom and good Anchorage, and is shelterd from all winds except those that blow from the North-East Quarter. It lies in the Latitude of 38 degrees 22 minutes South, and 4½ Leagues to the Northward of Gable end Foreland. Off the South point lies a small but high Island, so near to the Main as not to be distinguished from it. Close to the North end of this Island, at the Entrance into the Bay, are 2 high Rocks; one is high and round like a Corn Stack, but the other is long with holes thro' it like the Arches of a Bridge. Within these rocks is the Cove, where we cut wood and fill'd our Water. Off the North point of the Bay is a pretty high rocky Island, and about a Mile without it are some rocks and breakers. The variation of the Compass is here 14 degrees 31 minutes East, and the Tide flows at full and change of the Moon about 6 o'Clock, and rises and falls upon a Perpendicular 5 or 6 feet, but wether the flood comes from the Southward or Northward I have not been able to determine.[6]

When considered in the context of one of the most far-reaching exploratory expeditions in maritime history, this degree of fine detail for

5. Banks' Journal, 22 January 1770.
6. Cook's Journal, 29 October 1769.

what would have been to Cook a fairly nondescript location in a sparsely populated territory is indicative of how exacting he was on himself.

From the outset, Cook was determined that his map of New Zealand would be more than an outline of its coast. Not only was he and several others on board accumulating an abundance of intelligence about the territory – its fauna, flora, geology and inhabitants – he also wanted to construct a physical map of the landscape, showing mountains, rivers, lakes and, where possible, different elevations of land. Throughout his journal, Cook repeatedly mentioned his observations of the terrain further inland from the coast. These usually tended to be brief and general, principally because they were based on views from a distance (inland exploration was too dangerous and time-consuming on most occasions). One typical example is the comment he wrote when looking inland from Tauranga harbour in November 1769, when he recorded that 'the land Trends North-West as far as we could see, and appeared to be very rugged and hilly'. Less often, Cook would provide additional details and even a name for a particular geographic feature, as he did for the 820-metre-high Mount Edgecumbe, which lay 25 kilometres inland from the coast of Bay of Plenty:

> South-West by South from this Island on the Main land, seemingly at no great distance from the Sea, is a high round Mountain, which I have named Mount Edgcombe. It stands in the middle of a large Plain, which make it the more Conspicuous. Latitude 37 degrees 59 minutes South, Longitude 183 degrees 07 minutes West.[7]

These entries were then able to be used when Cook began to piece together his chart for the relevant portion of New Zealand that he was mapping at the time. The fact that most of the terrain that Cook attempted to depict on his maps was unreachable for him was frustrating, but something that he had to accept. For this disciple of accuracy, the vagueness of what lay inland left him portraying the topography in a fairly generalised fashion. One consequence of this was that the chaos of a great deal of the terrain that Cook saw appeared much more ordered and tidy on his maps. The fact that these maps were eventually produced either in black or white or with a few weak tones of ink-wash had the effect of further deadening the drama of the landscape and rendering it as something more geographically neutered. Also perhaps, from the

7. Ibid., 2, 3 November 1769.

perspective of prospective colonisers back in Britain, this vague view of New Zealand's topography, with some of the mystery of the unknown stripped away, made the location just that bit more enticing.

On top of so much else, the Admiralty required Cook to extend Britain's imperial boundaries in certain territories. He was instructed: 'with the Consent of the Natives to take Possession of Convenient Situations in the Country in the Name of the King of Great Britain: Or: if you find the Country uninhabited take Possession for his Majesty by setting up Proper Marks and Inscriptions, as first discoverers and possessors.'[8] It was a pretty ham-fisted approach to empire-building, even for this period, and it is unclear how seriously Cook took this injunction. Likewise, the Admiralty itself cannot have been overly committed to this expansionist policy, because, while it was content to have Cook claim territories in the name of the king, it was neither prepared nor apparently willing to act on such claims. There were to be no treaties, no civil administrations, no occupying forces, no legal jurisdiction and no boundaries of British authority imposed on the country. Instead, Cook would act out these faux annexations almost as a piece of imperial theatre. An example of such a performance occurred in mid-November 1769, while the *Endeavour* was anchored in Mercury Bay. Just before departing to continue the expedition northward, Cook recorded how he 'cut out upon one of the Trees near the Watering Place the Ship's Name, date, etc., and, after displaying the English Colours, I took formal possession of the place in the Name of His Majesty'.[9] Having done this, he gave not the slightest indication that such an act had any substance whatsoever beyond the gesture itself, while, for Banks, this pantomime of possession did not even warrant an entry in his journal. Clearly, the attitude of those on board the ship towards this aspect of the Admiralty's instructions was one of supreme indifference.

Of much more importance to Cook was the relentless process of mapping New Zealand with as much precision as he could muster. As he tacked northward from Mercury Bay, even small protrusions of rock rising just above sea level were recorded. Most of these did not appear on the final versions of his maps, but their precise locations were recorded in his journals, and so could serve as an aide for future mariners sailing in the vicinity.

Much of the work of Cook and his crew was pedestrian. Marking time, sounding depths, fixing positions and triangulating all the data to determine the exact form of the land that they were skirting had

8. Admiralty Commissioners, 'Secret. Additional Instructions', 30 July 1768.
9. Cook's Journal, 15 November 1769.

become routine. However, portions of the terrain sometimes presented specific challenges. On 18 October, the *Endeavour* rounded Cape Colville and headed towards the Firth of Thames. As the waters got shallower, and the weather remained 'thick and hazey',[10] Cook decided not to proceed further in the ship and, instead, opted to continue exploring on one of the small boats that was lowered from the side of the *Endeavour* with the captain and a handful of assistants. As they rowed through the gloom, they maintained sight of the coast (not having any idea of how far away the opposing coast might be), with Cook constantly sketching details along the way. This was the soon-to-be-named Coromandel Peninsula, which was relatively low-lying, the terrain tightly choked with ferns, from which ancient kauri trees (*Agathis australis*) – some probably over a thousand years old – stood out like arboreal sentinels above the fog.

Gradually, the mists lifted, enabling the *Endeavour* to move closer to the mouth of the river that emptied into this inlet. In the following days, Cook made use of the boat to explore areas close to the shore (fearful that the ship might get stuck on the muddy sea bed). The conditions on a small boat were far from ideal for charting the coast – it was slow and rocked more readily with each wave that lapped against it. Cook did his utmost to gather whatever measurements he could and then, once back in his cabin on the *Endeavour*, used them to complete his draft charts for the area. It was one of the slowest periods of mapping on the expedition but necessary to ensure that a comprehensive depiction of the coastline and surrounding topography could be assembled.

Having completed his cartographical survey of this portion of the Firth of Thames, Cook once more sailed north and, on 24 November, entered the Hauraki Gulf, where he had a rare lapse in attention. Possibly because he felt a mounting sense of urgency to get to the northernmost point of the territory, Cook bypassed a detailed survey of coast around what would become Auckland. Instead of his usual pedantic surveying and measuring of the coast, he almost casually noted in his journal that 'under the Western Shore lies some other Islands, and it appear'd very probable that these form'd some good Harbours likewise' and then added, as further evidence that this was not a priority for him, 'even supposing there were no Harbours about this River, it is good anchoring in every part of it where the depth of Water is Sufficient, being defended from the Sea by a Chain of Large and Small Islands which I have named Barrier Isles, lying aCross the Mouth of it'.[11]

10. Banks' Journal, 20 November 1769.
11. Cook's Journal, 24 November 1769.

The area from Ōrere Point to Cape Rodney – an approximately 160-kilometre length of coastline – was the only significant section of Cook's map of the North Island that was not drawn with a thick nib and cross-hatching. In its place, he drew what he estimated the shore to look like with a thin, faint line. This outline was sheer guesswork – and what makes it all the more baffling is that Cook had been so determined to explore the possibility of a strait located around 40 to 41 degrees latitude, yet he overlooked entirely the possibility of another strait in the Waitematā Harbour. In fact, there was a pinched section of land – no more than a kilometre wide – that Māori had used for centuries to haul their canoes from the east to the west coast of the island at this very location.[12] However, as Cook battled unfavourable winds while making his way northward, he made an educated guess about the shape of the coast in the distance – a guess that uncharacteristically turned out to be substantially incorrect.

For the remainder of the *Endeavour*'s passage northward, the mapping continued at roughly its normal pace, with Cook back to his fastidious charting. By the end of the month, the vessel had slipped into the Bay of Islands, which offered plenty of opportunity for exploration of the surrounding land. There were various encounters with members of local *hapū* and, as some indication of Cook's desire to be just rather than expedient, when he discovered three of his crew had stolen some food from a nearby Māori plantation, he punished each of them with a dozen lashes.[13]

After a week tracing the serpentine outline of the coast of the Bay of Islands, Cook decided it was time to resume the voyage north, even though, as he conceded:

> I have made no accurate Survey of this Bay; the time it would have requir'd to have done this discouraged me from attempting it; besides, I thought it quite Sufficient to be able to Affirm with Certainty that it affords a good Anchorage and every kind of refreshment for Shipping.[14]

This was an example of Cook's balancing act in practice. The coastal areas he mapped were done to a reasonable degree of detail but, aware that the

---

12. G. Irwin, D. Johns, R. Flay *et al.*, 'A Review of Archaeological Maori Canoes (*Waka*) Reveals Changes in Sailing Technology and Maritime Communications in Aotearoa/New Zealand, AD 1300-1800', *Journal of Pacific Archaeology* 8, no. 2 (2017), p. 39.
13. Cook's Journal, 30 November 1769.
14. Cook's Journal, 5 December 1769.

clock was ticking on his expedition, he compensated for the absence of a more meticulous charting of the coast by providing a summary of the benefits that the bay afforded future visitors by sea. Thus, satisfied with what he had accomplished at this spot, the *Endeavour* continued its cruise along the coast of the North Island, with gentle breezes, interspersed with periods of dead calm, governing its progress.

On 12 December, the wind picked up but from the wrong direction, forcing the ship to tack. Banks complained the following day that the wind was 'as foul as ever and rather overblows so that in this days turning we lost all we had gaind last week'.[15] After days spent zig-zagging close to the coast, Cook took the *Endeavour* almost 100 kilometres east in an effort to beat the weather – a seemingly counterintuitive decision, but one which seasoned sailors understood and which, on this occasion, was largely successful. The other advantage of this tactic was that Cook was able to judge the nature of the swells and the force and consistency of the winds, from which he was able to determine that there was no land north of them[16] and, therefore, that New Zealand terminated at the area that Tasman indicated, around Cape Maria van Diemen.

The weather was now only getting worse. By 14 December, the *Endeavour* was wrestling with wild winds and mountainous ocean swells. One of the sails tore when a sudden gust hit the ship and, for the first time since arriving in New Zealand, no land was in sight. The ship was blown off course and all Cook could do was wait until the storm passed over.

Two days later, with the worst of the squally weather having settled down, a speck of land around North Cape came into view for the crew member who was perched on the *Endeavour*'s mast. Over the next few days, Cook steered the ship around the northern extremity of New Zealand, approaching Cape Maria van Diemen, which had been the final location on the mainland that Tasman had named before sailing away from the territory. Cook provided a vivid depiction of the territory from the vantage point of the vessel, describing the tip of the North Island as:

> a peninsula juting out North-East about 2 Miles, and Terminates in a Bluff head which is flatt at Top. The Isthmus which joins this head to the Mainland is very low, on which account the land off the Cape from several situations makes like an Island. It appears still more remarkable when to the Southward of it by the appearance of a high round Island at the

15. Banks' Journal, 13 December 1769.
16. Cook's Journal, 14 December 1769.

> South-East Point of the Cape; but this is likewise a deception, being a round hill join'd to the Cape by a low, narrow neck of Land; on the South-East side of the Cape there appears to be anchorage, and where ships must be covered from South-East and North-West winds.[17]

As was becoming his habit when writing about New Zealand, Cook gave the sort of description (including locations that would offer shelter for ships) that he envisaged would be of benefit to later voyagers to this region. This complemented the information in his charts, even if there was no guarantee that such information would be included in the maps of the area that were eventually published.

After a brief respite from the violent weather that had pummelled the ship, in the final week of 1769, the crew of the *Endeavour* faced a storm of even greater severity. This was the worst weather of the entire voyage to date. Even some of the most experienced crew members on the ship went as far as to comment that they had never experienced such a ferocious storm in their lives.[18] By the beginning of January 1770, the conditions at sea had eased and Banks gave his own, more picturesque account of the west coast of the North Island as the *Endeavour* sailed towards its northern end:

> We venturd to go a little nearer the land which appeard on this side the cape much as it had done on the other, almost intirely occupied by vast sands: our Surveyors suppose the Cape shapd like a shoulder of mutton with the Knuckle placd inwards, where they say the land cannot be above 2 or 3 miles over and that here most probably in high winds the sea washes quite over the sands which in that place are low.[19]

There was no harbour in sight and so Cook continued to sail south, all the time keeping a watch on the barren, windswept, black-sand beach that could be seen extending interminably into a misty vanishing point in the far distance. It was an area where sky, sea and land all seemed to combine in the perpetual haze rising from the ocean breakers crashing onto the shore. Nowhere on the journey to date had anyone only and ever witnessed such desolation.

---

17. Cook's Journal, 19 December 1769.
18. Banks' Journal, 28 December 1769.
19. Ibid., 1 January 1770.

A Tempest and de Surville's Failed Attempt

The *Endeavour*'s route around the northern end of New Zealand had been one of the most torturous periods of the entire expedition. Most of the crew cowered below deck, inhaling the stench of body odour, vomit and accumulating urine and faecal matter as the hatches remained tightly closed, while the rain hammered the ship and the intense gales blew it far off course. On deck, it was just possible to hear the hoarse voices of those at the helm yelling to the few crew grappling with the rigging and hauling in the torn sails as they were battered by the elements. The sun and sky had disappeared behind a dense grey canopy of clouds, and the intense downpour reduced visibility to the point where all Cook could do was to wait out the storm, while praying that the ship would not be flung into any obstacles that might suddenly appear in these unknown waters. Such conditions confounded attempts at navigation. It was nearly impossible to get any bearings – either territorial or celestial – and measuring the distance that the *Endeavour* had been blown was reduced to little more than guesswork. Whatever trepidation Cook might have been experiencing, though, he kept it concealed. The conditions on the ship may have felt unbearable, but there was no alternative except to withstand the tempest until it eventually receded. It was at times like this that the utter isolation of the expedition was reinforced to everyone on board. The *Endeavour*'s crew was about as far away from European civilisation as it was possible to get. Or so they presumed.

* * *

In 1768, the recently constructed *Saint Jean-Baptiste* arrived in India. This was a vessel financed by a syndicate which included its captain, Jean-François Marie de Surville. De Surville had previously been employed by the French East India Company, but that organisation was on the verge of having its monopoly dissolved and becoming insolvent.[20] Seeing an opportunity to make some profit out of the financial rubble of the company, de Surville, together with some other shareholders, paid for the ship to be built, crewed and provisioned and to trade along the coast of India and what was then Ceylon. At around 600 tons, the *Saint Jean-Baptiste* was nearly twice the size of the *Endeavour* and carried a

20. E. Cross, 'The Last French East India Company in the Revolutionary Atlantic', *The William and Mary Quarterly* 77, no. 4 (2020), 613-40.

crew of 140 men.[21] It was a combination of the enormity of the vessel, along with a quest for new trading opportunities, and uncertainties over the future of commerce in the Indian subcontinent that spurred the syndicate that had funded the ship to look for a bigger return on their investment. However, exactly what would be the next step in this expansion was unclear. British maritime ascendancy in the region, and the corresponding French collapse, had narrowed the opportunities for traders outside the orbit of the British East India Company.

Desperate times – especially when facing financial ruin – can call for desperate decisions and so it was for de Surville. As the venture he had invested in looked like sinking, he was prepared to clutch at any opportunity that might save him and, in 1769, when financial salvation suddenly appeared in front of him, he grasped it tightly. In such circumstances, sound judgement quickly gave way to reckless conviction. In this case, it was a rumour doing the rounds in the Indian ports that de Surville was prepared to risk everything on.[22] Another Frenchman in the area at this time, the astronomer Abbé Alexis-Marie de Rochon, later jotted down some of the details of what had been said on the docks and in the drinking houses in the southern Indian ports regarding a tantalising new discovery by the British:

> I was in Pondicherry in August 1769 when the rumours spread that an English vessel had found in the South Sea a very rich island where, among other peculiarities, a colony of Jews had been settled. [T]he account of this discovery ... became so well known that it was believed in India that the purpose of Surville's voyage ... was to search for this marvellous land.[23]

This purportedly exotic locale, laden with gold and other riches (the reference to Jews having allegedly settled there was a standard type of anti-semitic allusion to the wealth that that place possessed) was in all

21. M. Lee, *Navigators and Naturalists: French Exploration of the New Zealand and the South Pacific Seas (1769-1824)* (Auckland: David Bateman, 2018), p. 40.
22. J.F.M. de Surville, *Extracts from Journals Relating to the Visit to New Zealand of the French Ship St Jean Baptiste in December 1769 Under the Command of J.F.M. de Surville*, trans. I. Ollivier and C. Hingley (Wellington: Alexander Turnbull Library Endowment Trust, 1982), p. 147.
23. A. Rochon, in J. Dunmore (ed.), *The Expedition of the* St Jean-Baptiste *to the Pacific, 1769-1770* (London: Hakluyt Society, 1981), p. 23, cited in Lee, *Navigators & Naturalists*, p. 41.

probability Tahiti, which Samuel Wallis had visited two years earlier and the location of which British authorities were still intent on keeping secret.[24] The source of this rumoured South Pacific Eldorado was most likely George Robertson,[25] the master of the ship that Wallis captained and a mapmaker who produced a chart of Tahiti.[26] It was the mysterious nature of this prized territory, together with the blind faith in immense wealth that only the truly desperate seem to possess, that gave this rumoured location an irresistible power of attraction.

Pierre Monneron, one of the officers who sailed with de Surville, described how their plan altered as soon as they heard about this island. With little care over the accuracy of the reports they were receiving, the commanders of the *Saint Jean-Baptiste* decided that 'this island was so extraordinary that it deserved the whole of their attention' and that 'they did not hesitate to arrange their equipment in order to prevent the English … from taking possession of the island'. A political dimension had now been added to the proposed venture to the South Seas, but the lure of profit was what would ultimately guide the expedition in that direction, as Monneron excitedly wrote in his journal:

> [E]ven allowing a good deal for exaggeration, it was quite natural to presume that the island must be much richer than any of the other countries, as it is situated about 700 leagues west of the Coast of Peru and in a southern latitude of 27 to 28 degrees, which is the latitude of Capiazo, where the Spaniards get gold from in immense quantities.[27]

Over the next few months, the ship was refitted and provisioned to the point where Monneron estimated it was ready for a voyage of up to three years. Nothing, he boasted, 'was neglected to put her in such a

24. E.C. Childs, *Vanishing Paradise: Art and Exoticism in Colonial Tahiti* (Berkeley: University of California Press, 2013), p. 22.
25. Robertson was the originator of the 'Jewish' suggestion that formed part of the rumour that the French picked up. H. Carrington (ed.), *The Discovery of Tahiti, A Journal of the Second Voyage of HMS Dolphin Round the World, under the Command of Captain Wallis, RN: In the Years 1766, 1767, and 1768, Written by her Master, George Robertson* (London: Hakluyt Society, 2017), p. 225.
26. R. Cock, 'Precursors of Cook: The Voyages of the Dolphin, 1764-8', *The Mariner's Mirror* 85, no. 1 (1999), pp. 36-37.
27. P. Monneron, in R. McNab (ed.), *Historical Records of New Zealand: Volume 2* (Wellington: John Mackay, 1914), p. 232.

state, as well as the crew, to endure the greatest fatigues'. However, while all this preparation was underway, the purpose of the expedition was kept firmly secret, with the French claiming that the vessel was about to be sent on a routine trading voyage to China and Batavia.[28]

On 2 June 1769, the *Saint Jean-Baptiste* sailed from India and, after threading its way through parts of Southeast Asia, headed south in the direction of New Zealand, which its crew sighted on 12 December 1769. De Surville initially planned to explore the coast extensively (relying on Bellin's map) but the deteriorating weather put these plans on hold, as the captain grappled with the challenging conditions:

> We continued to tack about, constantly looking for some port. The contrary winds lasted from the north until the 14th, then changed to the W.N.W. with such violence that several times we thought we would get wrecked. The sea was very rough, and constantly took us towards the land, which offered no accessible landing-place. ... We spent the whole of the night of the 14th until the 15th in great anxiety, obliged to tack about constantly, and to carry some sail, in order to drift less. The next day danger was just as great, the wind and the sea always being the same.[29]

Eventually, the violent winds eased off enough for de Surville to regain control of the vessel. He took measurements of his position and tentatively began to explore lengths of the coast in the far north of New Zealand. However, just when conditions seemed to be improving, the great storm that Cook was contending with in the final week of 1769 made itself felt to the French as well. Monneron noted the brutal effects of the weather, with the winds rising to the point where the ship's anchor cable snapped, causing the vessel to be tossed dangerously close to the shore. 'One cannot see death nearer than we did. We were very near the rocks, about 20 yards away', he wrote soon afterwards. The rudder bar broke in the storm, as did some of the rigging, and it was only through the 'greatest luck' that the crew and the vessel survived the elements.[30] It was during that week that the *Endeavour* and the *Saint Jean-Baptiste* sailed past each other. In ideal conditions, it is just possible that the

28. Ibid., p. 233.
29. Ibid., p. 265.
30. Ibid., p. 273

crews of the two ships might have caught a glimpse of each other[31] but, given the state of the weather that members of both ships so vividly described, neither had any chance of knowing how close they were.

It was also the weather that played a key role in de Surville's decision to depart for calmer waters. 'In a country so much exposed to storms as New Zealand is', Monneron explained, 'we could not possibly expose ourselves by remaining there longer. ... With but one anchor and a crew tired and reduced by half, could we without extreme imprudence decide to go back on our tracks?'[32]

During this short and interrupted visit, the French at least made some efforts to map portions of the coast, although these were unavoidably fragmentary.[33] One such example was the chart depicting Lauriston Bay (which Cook had recently named Doubtless Bay and which, incidentally, was where Kupe – the legendary Māori discoverer of New Zealand – initially made landfall in the country). The shore was plotted by the French with considerable accuracy, with some measurements included, along with a key denoting various land features in names that de Surville gave to the area.[34] However, time was against the French expedition and so there was never any chance of a comprehensive French map of the whole country being produced. Only four coastal charts outlining parts of northern New Zealand are known to have survived from the French expedition, along with various composite maps that were later produced.[35]

Subsequent circumstances also meant that the small amount of mapping of New Zealand undertaken by these French explorers would not eventually contribute to the first full map of the territory. Four months after sailing away from the country, de Surville drowned off the coast of Peru and his scurvy-ridden crew eventually returned to France in August 1773; only 66 of the original 173 men on board survived the

31. Lee, *Navigators & Naturalists*, p. 60.
32. M. Monneron, in McNab, *Historical Records of New Zealand: Volume 2*, p. 280.
33. Ibid., p. 287.
34. *Plan of the Bay of Lauriston, on New Zealand in 34°58'S, from a French M.S., December 1769* (London: A. Dalrymple [for the Admiralty Sept 20th], 1781), National Library of New Zealand, Alexander Turnbull Library Neg. 146966 1/2 and Neg. C1402.
35. J.R.H. Spencer, 'The Visit of the *St Jean-Baptiste* Expedition to Doubtless Bay, New Zealand, 17-31 December 1769', *Cartography* 14, no. 2 (1985), pp. 132-33.

expedition[36] (and any hope of the anticipated commercial spoils of the voyage had been completely abandoned). The venture was later fairly described as 'one of the most unfortunate and frustrated voyages in the annals of Pacific exploration by Europeans'.[37]

However, six years later, having returned from his own explorations of the region, Cook paid tribute to the cartographical work that de Surville had carried out:

> These Voyages of the French, tho' undertaken by private Adventurers, have been productive of some usefull discoveries, as well as contributing in exploaring the Southern Oceans. That of Captain Suville [*sic*] clears up a Misstake, which I was laid into by immaging the Shoals off the West end of New Caledonia to extend to the West as far as New Holland: it proves that there is an open Sea in this space and that we saw the NW extremity of that country.[38]

In addition to giving out praise where it was deserved, what this extract also reveals is that Cook was a studious scholar of the works of other cartographers and mapmakers. No map would be left unfolded as he worked endlessly to perfect his own charts.

36. Dunmore, *The Expedition of the* St Jean-Baptiste *to the Pacific, 1769-1770*, p. 125.
37. R.R.D. Milligan, 'Ranginui, Captive Chief of Doubtless Bay, 1769', *The Journal of the Polynesian Society* 67, no. 3 (1958), p. 182.
38. J. Cook, 'Revised holograph manuscript journal, *HMS Resolution* (1775)', British Library, Department of Manuscripts, additional manuscript 27888, p. 348.

Chapter 10

North and South Islands Revealed

Cook's Circumnavigation of South Island

Oblivious to his near encounter with a French vessel, Cook continued sailing along the west coast of the North Island. After passing Cape Maria van Diemen, he recorded on 2 January 1770 that its location was 'Latitude 34 degrees 30 minutes South; Longitude 187 degrees 18 minutes West from Greenwich'.[1] (The Greenwich meridian line was used to indicate 0° longitude and therefore served as a standard basis for measuring longitude.[2]) One of Cook's biographers noted that this measurement was 'extraordinarily accurate, seeing that the ship was never close to the Cape, and the observations were all taken in bad weather. The latitude is exact, and the longitude is only three miles in error.'[3]

Two days later, the *Endeavour* had reached the opening of Kaipara Harbour. An underwhelmed Cook wrote '[n]othing is to be seen but long sand Hills, with hardly any Green thing upon them, and the great Sea which the prevailing Westerly winds impell upon the Shore must render this a very Dangerous Coast'. Such was his aversion to this desolate location that he was determined never to visit it again if he could possibly avoid it.[4] He named this portion of the country 'The Desert Coast' and, as he continued sailing in a southerly direction, the large sand dunes

1. Cook's Journal, 2 January 1770.
2. W.G. Perrin, 'The Prime Meridian', *The Mariner's Mirror* 13, no. 2 (1927), 109-124.
3. W. Wharton, in Beaglehole, ed. *Cook's Journal*, 2 January 1770, p. 42.
4. Ibid., 4 January 1770.

in certain areas, along with the risk of getting too close to the shore, resulted in his map of this part of the North Island containing very few topographical features.

Further south, Cook was able to make out more detail in the immediate hinterland. On 10 January, the *Endeavour* sailed around 20 kilometres from the coast and in the evening came 'abreast of a Point of Land which rises sloping from the Sea to a Considerable height. … I named it Woodyhead.'[5] This was Mount Karioi, which Tasman had also seen, recorded and mapped. Nineteen kilometres offshore was a small rocky outcrop which Cook called Gannet Island for self-evident reasons, as he did when designating a craggy protrusion on the southern entrance of Kawhia Harbour Albatross Point.

By 13 January, the *Endeavour* had travelled further south to what would become known as New Plymouth. That evening, Cook wrote of seeing:

> the top of the Peaked Mountain to the Southward above the Clouds. … It is of a prodidgious height and its Top is cover'd with Everlasting Snow. … I have named it Mount Egmont in honour of the Earl of Egmont. … This mountain seems to have a pretty large base and to rise with a Gradual Ascent to the Peak, and what makes it more Conspicuous is its being situated near the Sea and in the Midst of a flat Country which afforded a very good Aspect, being Cloathed with Woods and Verdure.[6]

While this peak, known as Mount Taranaki, was indeed impressive (Banks described it as 'certainly the noblest hill I have ever seen'[7]), Cook paid just as much attention to the surrounding terrain, which he could see was relatively flat. Further inland, however, no one on the ship could know for certain what the country's topography was like and so Cook resorted to speculation, populating the blank spaces on his maps with a random sequence of hills. Their appearance signified that, as far as Cook could ascertain, the land was probably broken, hilly terrain, unlike the plane around Mount Egmont/Taranaki.

* * *

5. Ibid., 10 January 1770.
6. Ibid., 13 January 1770.
7. Banks' Journal, 13 January 1770.

Another piece of New Zealand's puzzle began to be solved by Cook from the middle of January 1770. In the first week that the *Endeavour* arrived in the country, Cook had detoured from his planned route in order to discover whether there was a passage leading from the East Coast to what Tasman had named Murderers' Bay. However, having been confronted by nothing but coast at the relevant latitude, he proceeded to sail north – temporarily abandoning this line of investigation. Now, though, Cook was once again approaching the same latitude, but on the west of the island, where Tasman had carried out his own cursory exploration in 1642. On 14 January 1770, the northern end of the South Island came into view, although, at this point, no one on board the ship had any way of knowing that it was a separate landmass from the one whose coast they had been following. So far, all the evidence pointed to the probability that they were about to enter a large bay, of the sort that had been depicted on maps of New Zealand up until this time.

For the next several days, Cook cautiously threaded the *Endeavour* through Queen Charlotte Sound, which was known to local Māori as Tōtaranui. He soon came to like the area because of the ready availability of provisions and the considerable cooperation he received from the indigenous inhabitants. By this stage in the voyage, Cook had developed a reliable idea of the coastal length of the North Island, and how long it had taken to circumnavigate it almost completely. Looking at Tasman's map, in mid-January 1770, he was now in a position to make an informed estimation of the remaining landmass of the territory. Yes, there were two major unknowns: whether the area he was now in was connected to the territory he had explored so far or separated by water; and how far south the landmass extended beyond what Tasman had drawn. Cook's intuition was by this stage beginning to guide him in coming to some conclusion about both of these unknowns. He surmised, first, that the remaining territory of New Zealand would be greater than that which he had met so far (thus making him even more conscious of how little time he had to carry out a full exploration of the country) and, second, that there must be some passage separating what he would soon identify as the North Island from the territory that lay to the South.

The problem for Cook was that, if he was wrong in his presumption about a passageway separating the two landmasses that he supposed constituted New Zealand, then he ran the risk of consuming potentially weeks mapping what could turn out simply to be a vast, deeply indented bay. He needed an aerial view of the region to give him some sense of what might lie ahead and, on 22 January, he achieved this. Accompanied by one of his crew, he landed on the southern end of Arapaoa Island (the easternmost island in Queen Charlotte Sound). As soon as they had dragged their boat up onto the beach, the two men began to look for any elevated point that might give them a good outlook of the area. After initially scaling one hill, only to find the view at the top obscured by dense forest, they eventually managed to ascend a peak known as Kaitapeha, where Cook got the reward he was after. Standing on this exposed promontory, about 380 metres above the point where they had landed their boat, Cook wrote later that day:

> I saw what I took to be the Eastern Sea, and a Strait or passage from it into the Western Sea; a little to the Eastward of the Entrance of the inlet in which we now lay with the Ship. The Main land which lies on the South-East side of this inlet appeared to me to be a narrow ridge of very high hills, and to form a part of the South-West side of the Strait; the land on the opposite side seem'd to tend away East, as far as the Eye could see. To the South-East appeared an Open Sea, and this I took to be the Eastern.[8]

It was no wonder that he returned to the *Endeavour* 'in high spirits',[9] having confirmed the existence of what from 1770 was known as 'Cook's Straight' (and, later, Cook Strait).

On the final day of January, Cook once more reverted to formalities and, in deference to his instructions, hoisted the Union flag and took possession of Queen Charlotte Sound in the name of the king. An elderly Māori was present during the ceremony and, with Tupaia acting as translator, Cook quizzed him about the area and, in particular, the strait that Cook had seen leading eastwards to the Pacific Ocean. '[H]e very plainly told us there was a Passage', Cook recorded, 'and as I had some Conjectures that the lands to the South-West of this Strait (which we are now at) was an Island, and not a Continent, we questioned the old Man about it.' The outcome of this conversation revealed that Māori had developed some conception of the form of New Zealand. This aged informant revealed that the country:

> consisted of two Wannuas, that is 2 lands or Islands that might be Circumnavigated in a few days, even in 4. This man

8. Cook's Journal, 23 January 1770.
9. Banks' Journal, 22 January 1770.

> spoke of 3 lands, the 2 above mentioned which he called Tovypoinammu which Signifies green … Stone, such as they make their Tools or ornaments, etc., and for the third he pointed to the land on the East side of the Strait; this, he said, was a large land, and that it would take up a great many Moons to sail round it; this he called Aeheino Mouwe, a name many others before had called it by. That part which borders on the strait he called Teiria Whitte.[10]

Some of the confusion about circumnavigating an island in a few days most likely arose from mistranslation, but what is abundantly clear from this passage is that Māori knowledge of the country's geography extended beyond the immediate vicinity where each community lived. This also hints at a complex and intricate network of information exchange between various tribes probably extending back over several centuries, in which intelligence about the country's topography was accumulated and refined. It is therefore extremely odd that the inherently curious Cook did not appear to give any thought as to how Māori had acquired this knowledge, given the absence of any instruments of measurement, and the fact that all such information had to be stored in the memory and communicated orally, as this was a preliterate society.

* * *

With a light breeze carrying the *Endeavour* out of Queen Charlotte Sound, in early February 1770, Cook commenced his passage through Cook Strait, narrowly avoiding crashing onto a cluster of islands – known as the Brothers – as he headed south-east. The ship followed the coast of the North Island past Whanganui-a-Tara (later Wellington Harbour and Palliser Bay) and, at eleven o'clock on the morning of Friday, 9 February, Cook came within sight of Cape Turnagain, thus completing his circumnavigation of the North Island. Immediately the location was confirmed, 'we once more turnd our heads to the southward', as Banks recorded triumphantly.[11]

The *Endeavour* had earlier sailed by the northernmost parts of the South Island, but now it was Cook's duty to chart that landmass with the same thoroughness that he had applied to the North Island. The remaining question which overshadowed all others was whether this

10. Cook's Journal, 31 January 1770.
11. Banks' Journal, 9 February 1770.

was indeed another island – as Māori from Queen Charlotte Sound had informed him – or whether it was a colossal protuberance jutting out from the fabled Great Southern Continent. Five days after turning around at Cape Turnagain, the *Endeavour* was off Kaikoura, on the eastern coast of the South Island and, by 17 February, Cook mapped an island 'which I have named after Mr. Banks'[12] but which was, in fact, a peninsula, which survives with Banks' name attached to it today. Believing this to be an island was an elemental error which can be put down to the haste with which Cook was now working. He was maintaining the routine of measuring distances sailed and positions reached but was obviously conscious that he did not have enough time to explore every cavity in the coast that he came across.

Two days' sailing from this point, Banks began to speculate about the territory Cook and others were involved in mapping. Looking at the land rising far to the west, the botanist:

> once more cherishd strong hopes that we had at last compleated our wishes and that this was absolutely a part of the Southern continent; especialy as we had seen a hint thrown out in some books that the Duch, not contented with Tasmans discoveries, had afterwards sent other ships who took the land upon the same lat. as he made it in and followd it to the Southward as high as Lat 64°S.[13]

When it was soon discovered this land was an island, little thought was given by most of the crew from that point onwards about the Great Southern Continent but, in mid-February 1770, it was still very much a viable possibility for Banks, at least, who was almost certain that it existed. However, he was very much in the minority among those on the *Endeavour*, as he conceded a few days later:

> We were now on board of two parties, one who wishd that the land in sight might, the other that it might not be a continent: myself have always been most firm for the former, tho sorry I am to say that in the ship my party is so small that I firmly beleive that there are no more heartily of it than myself and one poor midshipman.[14]

---

12. Cook's Journal, 17 February 1770.
13. Banks' Journal, 19 February 1770.
14. Ibid., 24 February 1770.

The voyage continued heading in a southerly direction and, despite diversions brought on by bad weather, in early March, it reached the southern tip of the South Island. Cook's chart outlined part of what he thought 'has very much the appearance of an Island' (Stewart Island) and described it as having 'a very barren Aspect; not a Tree to be seen upon it, only a few Small Shrubs'.[15] However, feeling more pressed for time than had been the case earlier on in the expedition, Cook dismissed this possibility and determined that the land in question was probably a peninsula extending from the southern tail of the South Island. Parkinson, though, had his doubts. The land which he saw from a distance 'seemed to be an island, having a great opening between it and the land which we had passed before; but the captain designing to go round, we steered for the south point hoping it was the last'.[16] This was a rare instance when Cook's haste led to a misrepresentation on the New Zealand map that he was compiling.

By the middle of the month, the *Endeavour* was on the home stretch, sailing north past Taiari Inlet and along the Fiordland coast. This glacier-carved section of the South Island was ripe for exploration but Cook opted to avoid charting its detailed indentations and, instead, produced a depiction of the coast that almost omitted the fiords altogether, offering readers the impression of slight corrugations where there were major incisions. Part of the reason for this was Cook's rising sense of impatience with the expedition, but there were also the particular practicalities involved in undertaking a detailed exploration of this area. On 14 March, Cook explained precisely what the problem was as he approached the mouth of one of these fiords:

> The land on each side ... riseth almost perpendicular from the Sea to a very considerable Height; and this was the reason why I did not attempt to go in with the Ship, because I saw clearly that no winds could blow there but what was right in or right out, that is, Westerly or Easterly; and it certainly would have been highly imprudent in me to have put into a place where we could not have got out. ... I mention this because there was some on board that wanted me to harbour

15. Cook's Journal, 9 March 1770.
16. R. McNab, *Murihiku and the Southern Islands: A History of the West Coast Sounds, Foveaux Strait, Stewart Island, the Snares, Bounty, Antipodes, Auckland, Campbell and Macquarie Islands from 1770 to 1829* (Invercargill: William Smith, 1907), p. 5.

> at any rate, without in the least Considering either the present or future Consequences.[17]

This final comment is instructive. While Cook was guided by practicality, obviously, some of those on board were inclined to succumb to their curiosity. However, it was the captain's decision what course the ship took and, at this time, there was just one thought in Cook's mind: keep sailing north.

At some time during 18 March, the *Endeavour* passed that anonymous point on the coast between Ōkārito and Hokianga where Tasman had first encountered the country. Neither Cook nor the even more sentimentally inclined Banks made any mention of this, suggesting that by this stage in the voyage, Tasman's map was no longer being referred to, at least on a regular basis.

With indifferent weather, the *Endeavour* ploughed its way through the swells, reaching Cape Farewell on 23 March. Cook then steered due east towards Stephens Island, which he recognised from when the expedition had been in Queen Charlotte Sound in January. Over the next few days, Cook undertook little in the way of mapping and, instead, focussed on provisioning the ship for the next leg of the voyage (westwards to the land that he would claim for Britain as New South Wales). On 31 March, he wrote that he was 'now resolv'd to quit this Country altogether, and to bend my thought towards returning home by such a rout as might Conduce most to the Advantage of the Service I am upon'. He convened a meeting with the *Endeavour*'s senior officers on how to proceed and proposed that they return home via Cape Horn, which would enable them to establish definitively whether the Great Southern Continent existed, even though, as he conceded, the ship by this stage was not in good condition,[18] with its sails and rigging in a particularly bad state.

As the crew made the final preparations to depart New Zealand, Banks sat down in his cabin to record a more thorough account of his visit to the country and some concluding comments on its form:

> Having now intirely circumnavigated New Zealand and found it, not as generaly has been supposd part of a continent, but 2 Islands: and having not the least reason to imagine that any countrey larger than itself lays in its neighbourhood, it was resolvd to leave it and Proceed upon farther discoveries in our

17. Cook's Journal, 14 March 1770.
18. Ibid., 31 March 1770.

> return to England being determind to do as much as the state of the Ship and provisions would allow ... although we hopd to make discoveries more interesting to trade at least than any we had yet made, we were obligd intirely to give up our first grand object, the Southern Continent: this for my own part I confess I could not do without much regret. – That a Southern Continent realy exists, I firmly beleive; but if ask'd why I beleive so, I confess my reasons are weak; yet I have a preposession in favour of the fact which I find it dificult to account for.[19]

Cook, on the other hand, had few doubts that the Great Southern Continent was little more than groundless supposition. The circumnavigation of New Zealand had delivered a sizeable blow to the existence of this fabled continent and, although Banks continued to have faith in the idea, his more pragmatic captain thought otherwise.

The scale of the work that Cook and his crew undertook during the New Zealand stage of the expedition, when peeled away to its statistical core, gives a sense of the monumental dimensions of this undertaking. He took 117 days to map around 3,800 kilometres of the country's coastline (sailing an average of 32 kilometres a day) and spent a total of almost two months anchored in harbours and other places of shelter, where further observations were conducted and where encounters with various Māori communities took place.[20]

The New Zealand portion of this expedition was now complete and Cook had a comprehensive set of charts, maps, written accounts and sketches depicting the territory in what was, for this era, extraordinary precision. The French mariner Julien-Marie Crozet (who accompanied Marc-Joseph Marion du Fresne on a voyage to New Zealand in 1772-73) wrote that, as soon as he obtained a copy of Cook's charts, he compared it to the mapping that the French had undertaken. His verdict left no room for doubt; he described Cook's maps as being 'of an exactitude and of a thoroughness of detail which astonished me beyond all powers of expression, and I doubt much whether the charts of our own French coasts are laid down with greater precision'.[21] Yet, in spite of his

19. Banks' Journal, 31 March 1770.
20. A. David (ed.), *The Charts and Coastal Views of Captain Cook's Voyages: Volume 1: The Voyage of the* Endeavour, *1768-1771* (London: Hakluyt Society, 1988), p. xxxii.
21. J.-M. Crozet, in ibid., p. xxxiv.

unrivalled expertise, Cook refused to allow himself to be cocooned by his own genius and was open to information from other sources. The naturalists on board certainly contributed to the broader process of depicting the place is being mapped, while several of the officers were themselves highly specialised in obtaining the various types of measurements necessary for Cook to undertake his work. Moreover, unlike most previous European visitors to the South Pacific, Cook was also open to receiving indigenous intelligence relating to navigation and mapping. There was a hint of this with the elderly Māori who had informed them on 31 January that New Zealand was not a single landmass. Such a profound geographical awareness – made all the more intriguing and even puzzling by the fact that such knowledge was developed, stored and transmitted solely in oral form – serves as a thin shaft of light illuminating a small patch of what must have been an extensive body of indigenous knowledge about the geography of the region.

Tupaia

It almost might never have happened. On 3 May 1769, while Banks and Solander were walking inland from the shore in Tahiti, by chance, they unknowingly stepped into a different intellectual universe. It was there that they encountered Tupaia, a native of Ra'iātea (the second largest of the Society Islands), who happened to be present in the vicinity. Beneath the commonplace surface of the island were a select few people who possessed a rich seam of expertise in particular fields. In Tupaia's case, it was a specialised knowledge of navigation and geography that both bewildered and impressed these European visitors.

Initially, though, Tupaia's skills met with an ambivalent response from some of those on the *Endeavour*. Consternation at his insights into the region's physiography fluctuated with disdain at the possibility that someone from such a 'primitive' environment could possibly be so well-informed without having obtained his knowledge from a European source. Tupaia's brilliance was counter-intuitive to those for whom European civilisation represented the apex of human accomplishment. Yet, his expertise was undeniable and, within a few years, he had come to be regarded by those acquainted with his work as a 'genius'.[22] The

22. G. Forster, *A Voyage Round the World, in His Britannic Majesty's Sloop, Resolution, Commanded by Capt. James Cook, during the Years 1772, 3, 4, and 5*, 2 vols (London: G. Robinson, 1777), Vol. 1, p. xvi.

Irish seaman John Marra, who joined the *Endeavour*'s crew at Batavia, shortly after Tupaia had died (on 20 December 1770), became familiar with the traces of Polynesian navigation and cartography that Tupaia had bequeathed to the expedition. His appreciation of this type of knowledge was made clear in his journal (which was published in an edited form five years later):

> As their [Polynesian] whole art of navigation depends upon their minutely observing the motions of the heavenly bodies, it is astonishing with what exactness their navigators can describe the motions and changes of those luminaries. There was not a star in their hemisphere fixt or erratic but Toobia [Tupaia] could give a name to, tell when and where it would appear and disappear; and what was still more wonderful, could foretell from the aspect of the heavens the changes of the wind, and the alterations of the weather, several days before they happened. By this intelligence he had been enabled to visit most of the islands for many degrees round that of which he was a native. By the sun they steer in the day, and by the stars they steer in the night; and by their skill in presaging the weather, they can without danger lengthen or shorten their voyage as appearances are for or against them.[23]

Tupaia was an *arioi*[24] or *tahu'a* which translates loosely as priest, but which was a role that encompassed esoteric functions as well as specialist temporal skills. Initially, Cook refused to take Tupaia on board as an extra passenger – doubting that there was any value to be had in him joining the voyage, but Banks pushed for his inclusion and offered to take full financial responsibility for this guest. Banks was effusive about Tupaia's value both to the expedition and to him personally. 'He is certainly a most proper man', the botanist wrote:

> well born, chief ... or preist of this Island, consequently skilld in the mysteries of their religion; but what makes him more than any thing else desireable is his experience in the navigation of these people and knowledge of the Islands in

23. J. Marra, *Journal of the Resolution's Voyage: in 1772, 1773, 1774, and 1775* (Dublin: Caleb Jenkin, 1776), p. 217.
24. Salmond, *Aphrodite's Island*, pp. 36-37, 203-4.

> these seas; he has told us the names of above 70, the most of which he has himself been at.[25]

Either Banks had misunderstood Tupaia – a distinct possibility given the enormous language barrier separating the two – or Tupaia had exaggerated the breadth of his geographical knowledge, but, regardless, on the basis of such an impressive claim, Banks succeeded in persuading Cook that Tupaia should join the expedition.

Cook's earlier cynicism about Tupaia's skills eventually dissipated. After observing him over the course of several days, Cook 'found him to be a very intelligent person, and to know more of the Geography of the Islands situated in these Seas, their produce, and the religion, laws, and Customs of the inhabitants, than any one we had met with'.[26] Yet, doubts remained. While Banks had boasted that Tupaia had sailed to around 70 islands, Cook initially regarded Tupaia's descriptions as 'Vague and uncertain' and concluded that this indigenous navigator had probably only visited around twelve of the locations he mentioned.[27]

Banks' view of Tupaia began to shift as well in the first week of his acquaintance with him, but in a far more derogatory direction; he wrote of his desire to 'keep' Tupaia as a sort of Polynesian pet.[28] By the time the expedition reached the shores of New Zealand, however, Tupaia's reputation among everyone on the vessel had altered. In particular, his incalculable value as a translator was unquestioned by those on the *Endeavour*, and this aspect of his expertise seems to have eclipsed his skills as a navigator and cartographer at the time. Nevertheless, it is his maps which are the most intriguing artefacts that Tupaia left, and which hint at the sort of skills and knowledge that had evolved in the South Pacific in this field, independently of any European influence.

The drafts of the maps that Tupaia created of Tahiti are lost, and there is only a faint clue as to their means of production. According to George Forster (who accompanied Cook on his second expedition to the southern Pacific and whose account is second-hand), once on board the *Endeavour*, Cook and others showed Tupaia the charts that they had recently produced[29] and 'having soon perceived the meaning and

25. Banks' Journal, 12 July 1769.
26. Cook's Journal, 13 July 1769.
27. Ibid.
28. Banks' Journal, 12 July 1769.
29. L. Eckstein and A. Schwarz, 'The Making of Tupaia's Map: A Story of the Extent and Mastery of Polynesian Navigation, Competing Systems of

use of charts, he [Tupaia] gave directions for making one according to his account'. The phrase 'gave directions' suggests that another hand was holding the pencil when the sketch maps were made, with Tupaia pointing to places on the paper and describing what to draw.[30]

Forster obtained his copy of Tupaia's maps from Banks, who, 'with great politeness and well known readiness to promote whatever has a tendency to become subservient to science, permitted me to take a copy of it'. Forster again noted that these 'original' works were 'drawn after Tupaya's direction' rather than directly in his own hand.[31] This was one of those cases of cultural coalescence involving the migration of this material from the Polynesian mind to a British archive, with Banks or Cook acting as intermediaries in transfiguring Tupaia's ideas into a European likeness. It is impossible to know what liberties they took in this process, but the surviving reproductions contain several elements (such as names in text, compass orientations, longitude coordinates, specific cross-hatching to signify coastlines, the form of topographical renditions of mountains, and images of European ships) that were obviously foreign embellishments of Tupaia's indigenous view.[32] The two versions of Tupaia's charts in Banks' collection were converted by Johann Reinhold Forster into a single chart, which he memorialised as a 'monument to the ingenuity and geographical knowledge of people of the Society Islands', and which he published in 1778.[33]

When the *Endeavour* sailed into New Zealand waters, Tupaia became the vital cultural and linguistic conduit between those on board the vessel and the various Māori communities that they visited. It is hard to imagine the expedition amassing even a fraction of the information it did about the country without his assistance. Although Tupaia did not contribute to Cook's cartographical work in New Zealand, the revelatory fragment that he did unveil (along with the intelligence that Cook received from

---

Wayfinding on James Cook's *Endeavour*, and the Invention of an Ingenious Cartographic System', *Journal of Pacific History* 54, no. 1 (2019), p. 21.

30. Forster, *A Voyage Round the World*, p. 511.
31. Ibid., p. 512.
32. Tupaia's Map, 1770, British Library, BL Add MS 21593.C (T3/B); Forster, *A Voyage Round the World*, p. 512 (insert); 'Chart of the Society Islands', possibly drawn by Joshua Banks in collaboration with Tupaia, British Library, BL Add MS 15508, f. 18.
33. R. Langdon, 'The European Ships of Tupaia's Chart: An Essay in Identification', *Journal of Pacific History* 15, no. 4 (1980), p. 225; A. Di Piazza and E. Pearthree, 'A New Reading of Tupaia's Chart', *The Journal of the Polynesian Society* 116, no. 3 (2007), p. 324.

Māori in Queen Charlotte Sound about the shape and dimensions of New Zealand) is more than sufficient to demonstrate that the navigators and cartographers on board the *Endeavour* were not the only specialists in these fields as they charted the South Pacific. Tupaia's full potential was never realised on this voyage but the echoes of his thinking, outlined on paper by Cook and Banks, give some impression of two very different navigational worlds just beginning to come into each other's orbits.

Chapter 11

The End of Cook's First Journey to the Southern Hemisphere

The Journey Home: Eastern Australia, Batavia and Britain

With the circumnavigation and charting of New Zealand complete, Cook rolled up his maps and sketch books and sailed west for Tasmania – the reverse route that Tasman (to whose journal he was once again referring) had taken. However, as he neared the island, a strong gale forced the ship northwards and, with neither instructions nor available time to investigate Tasmania, Cook satisfied himself with exploring the east coast of what is now mainland Australia (claiming what he named New South Wales for Britain), before making his way to Tasman's former base – Batavia – which the *Endeavour* reached in mid-October 1770.

As soon as he dropped anchor in the port, Cook immediately requested a report on the condition of the ship. The news was not good. The hull was seriously damaged in parts and the vessel was taking on water – a problem that was made worse because one of the pumps had become so worn that it was now 'useless' and the others were nearing the same condition.[1] The Dutch authorities were extremely cooperative and Cook soon was able to make arrangements for repairs to commence (even seeking finance from the Dutch to help cover the cost of overhauling the ship).

As work on the *Endeavour* commenced, Cook was informed that one of the Dutch fleets was about to depart for the Netherlands. He located

1. Cook's Journal, 10 October 1770.

the commander of the fleet, Captain Frederick Kelger of the *Kronenburg*, and handed him a packet addressed to Sir Philip Stephens, the First Secretary of the Admiralty, 'containing a Copy of my Journal, a Chart of the South Sea, another of New Zeeland, and one of the East Coast of New Holland'.[2] With a gruelling 18,000 kilometres still to be sailed by Cook before reaching England – including rounding the notoriously wild and cruelly misnamed Cape of Good Hope – despatching a copy of all his records to London was his insurance if the worst should happen and the *Endeavour* sank, losing two years of meticulous mapmaking in the process.

The copying process had most likely taken place simultaneously with the original version of Cook's journal being written. There is certainly little chance that this work – exceeding 170,000 words, was duplicated by hand in the fortnight after the expedition had arrived in Batavia, and there is no record of the officers sitting down to undertake such an arduous task after leaving New Zealand. Instead, Cook's clerk on the *Endeavour*, Richard Orton, regularly updated a copy of Cook's journal (with his own idiosyncratic spelling and punctuation conventions). The process that Cook relied on is useful in revealing part of his approach to mapmaking:

> Cook wrote up his log, sometimes at a fair length, as if it were a trial run for the journal, leaving blanks here and there for the names of bays or headlands not yet conferred, or bearings or positions not yet fully worked out. Then, still keeping on his left-hand pages the essentials of daily navigational detail, he wrote his journal proper, leaving similar blanks – for this journal was very close to the events. Sometimes, oddly enough, he later filled the blanks in the log and not in the journal.[3]

Yet, having achieved a feat of towering importance in the history of navigation and cartography, Cook let slip only a small measure of self-congratulation in an otherwise deferential and unassuming summary of his work which he appended to the copy of his journal and charts that he sent to England. 'Altho' the discoverys made in this Voyage are not great', he wrote with undue modesty:

2. Ibid., 24 October 1770.
3. J.C. Beaglehole, 'Some problems of editing Cook's journals', paper read to the history section of the Australian and New Zealand Association for the Advancement of Science (Dunedin, January 1957), p. 25.

> yet I flatter myself they are such as may Merit the Attention of their Lordships; and altho' I have failed in discovering the so much talked of Southern Continent (which perhaps do not exist), and which I myself had much at heart, yet I am confident that no part of the Failure of such discovery can be laid to my charge.[4]

The idea of the Great Southern Continent had never been an abiding interest of Cook's, and there is every indication that he thought its existence unlikely. It was his maps, however, that were his primary focus and a source of pride. 'The plans I have drawn of the places I have been at were made with all the Care and accuracy that time and Circumstances would admit of', he informed Stephens, adding that: 'I am certain that the Latitude and Longitude of few parts of the World are better settled than these.'[5]

No details have survived of what happened to the documents Cook despatched to London via the *Kronenburg*, but some elemental deductions can be made. First, there is no reason to presume that the lack of subsequent mention of the parcel is suggestive that it did not reach its destination. On the contrary, had the Admiralty later discovered that the Dutch had failed to hand over such an important consignment of intelligence, it is probable that this would have left some angry correspondence in the archives. What is more likely is that this spare set of records reached London after Cook had returned to the British capital and so became a redundant copy as far as the Admiralty was concerned. The voyage by Dutch ships from Batavia to Amsterdam at this time took around seven and a half months,[6] meaning that the *Endeavour* journals and maps would have arrived in the Netherlands around mid-June 1771. From there, it would have had to pass through the hands of various VOC bureaucrats, and then be sent on to London – a relatively short voyage, but one that was frequently delayed because of the adverse winds along the route.[7] It is therefore probable that Cook's records of the expedition reached the Admiralty before the package from Amsterdam was delivered to London.

* * *

4. Cook to P. Stephens, 23 October 1770, in Cook's Journal, 24 October 1770.
5. Ibid.
6. Bruijn, 'Between Batavia and the Cape', p. 265.
7. P. Koudijs, 'Those Who Know Most: Insider Trading in Eighteenth-Century Amsterdam', *Journal of Political Economy* 123, no. 6 (2015), p. 1357.

On 26 December 1770, the freshly repaired and fully replenished *Endeavour* departed Batavia and headed west towards southern Africa. Then, after rounding the Cape of Good Hope, Cook steered north towards home. On 4 May 1771, as the *Endeavour* was approaching Ascension Island in the South Atlantic, a British ship (HMS *Portland*) came into view. The two vessels signalled each other, and, on 10 May, the *Portland* drew up alongside, with its captain, John Elliot, coming on board the *Endeavour*. As the *Portland* was the faster of the two vessels, Cook decided to hand over to Elliot 'a Letter for the Admiralty, and a Box containing the Ship's Common Log Books, and some of the Officers' Journals, etc. I did this because it seem'd probable that the Portland would get home before us, as we sail much heavier than any of the Fleet.'[8] As it turned out, the *Portland* reached the English coast just three days before the *Endeavour*.

On 13 July, Cook made the final entry in his journal: 'At 3 o'clock in the P.M. anchor'd in the Downs, and soon after I landed in order to repair to London.'[9] Once the *Endeavour* had dropped anchor, a pilot took over the vessel and guided it along the coast and into the mouth of the Thames, docking at Woolwich (where the weary crew received their pay).[10] However, as soon as the ship had reached the Downs, near Dover, Cook went ashore and boarded the next available coach for the 113-kilometre trip along the route to London. The journey ended at London Bridge, where he alighted and then made his way to the Admiralty at Whitehall. After hanging on tightly to every aspect of the expedition for nearly three years, Cook had now let go. He briefly discussed with officials some of the details of the documents and charts relating to the voyage[11] – which, at the time, was the largest collection of records pertaining to a single expedition that the Admiralty had ever received – and then left the building to resume a short stint of civilian life.

What Happened to All the Journals, Maps and Charts?

Although the fact of Cook's expedition was widely known among British officials, politicians and the scientific community, aspects of its purpose and some of its destinations were still strictly secret as the *Endeavour*

---

8. Cook's Journal, 10 May 1771.
9. Ibid., 13 July 1771.
10. N. Erskine, 'The *Endeavour* after James Cook: The Forgotten Years, 1771-1778', *The Great Circle* 39, no. 1 (2017), p. 57.
11. David (ed.), *The Charts and Coastal Views of Captain Cook's Voyages*, p. xli.

set sail on its epic voyage in 1768. At the Admiralty's insistence, parts of the mission had been so clandestine that even Cook himself was not made privy to them until part-way through the journey. However, in the three years since these orders had been issued, there had been a sea change in Britain's approach to intelligence gathering and sharing. The Admiralty had come to realise (as the Dutch East India Company had in the previous century) that keeping the findings of such an expedition confidential served no strategic purpose. On the contrary, not only did the apparatus of secrecy require tiresome and expensive effort in pursuit of a futile objective (it was well known that the first leak on any voyage was of information that sailors shared in ports), but publicising the sorts of extraordinary findings that Cook had made could be turned into an exercise in brandishing British patriotism and imperial prowess.

Cook must have had more than an inkling that the Admiralty's insistence on confidentiality had been overcooked. After all, when he reached Batavia, he did not hesitate in placing a copy of the entire record of his journey to date, including charts, maps, journals and measurements, in the hands of a Dutch vessel (which would take this parcel initially to Amsterdam before it was sent on to London). If there had been any serious apprehension at all about the secrecy of the expedition being breached, Cook would not have even contemplated this course of action. He heard, while in Batavia, that the French had also been probing parts of the South Pacific and realised that there was probably little point in concealing details about his recent discoveries when anyone else exploring the region would be able to assemble similar information.

By the time that Cook returned to England, the decision had been made at the highest levels of the Admiralty to leap from secrecy to spectacle when it came to the question of how to deal with the mass of documents generated during the voyage. In an unusual move, a skilled author was to be employed by the government, along with engravers and artists as necessary, to produce a full published account of the expedition. It would be based primarily on Cook's journals and maps, augmented in parts with content from others who had sailed with him. Cook was fully supportive of such moves. At the close of his journal, he mentioned two Spanish ships that he had found out had visited Tahiti several months before the *Endeavour* arrived there, in addition to intelligence he had gathered on the French in the region. He even pointed out that:

> [he] was told by some French Officers, lately come from the Island Mauritius, that Orette, the Native of George's Island which Bougainville [a French captain] brought away with him,

> was now at the Maritius [*sic*], and that they were going to fit out a Ship to carry him to his Native country, where they intend to make a Settlement; 100 Troops for that purpose were to go out in the same Ship. This account is confirmed by a French Gentleman we have on board, who has very lately been at the Maritius [*sic*].[12]

Such plans, if they were indeed true, and not just the latest chatter to roll off the rumour mill, were never executed. However, although Cook may have believed such accounts, he was less concerned with their veracity than with the possibility that his own accomplishments – carried out in the name of Britain – were at risk of being seriously undermined. In his mind, he began to construct a scenario in which the French would approach the coast of New Zealand and sail in certain directions, mapping as they went. His principal worry was that they would:

> fall in with the most fertile part of that Country, and as they cannot know anything of the Endeavour's voyage, they will not hesitate a moment to declare themselves the first discoverers. Indeed, I cannot see how they can think otherwise. ... Thus this Island, though of little value, may prove a Bone of Contention between the 2 Nations, especially if the French make a Settlement upon it.[13]

The only solution that Cook could see for preventing such an eventuality was for the records of his voyage promptly to be 'published by Authority to fix the prior right of discovery beyond disputes'.[14] Banks shared these concerns and mentioned in a letter to a friend, written in December 1771, that on returning to England, he had 'put all the Papers relative to ye adventure of it into ye hands of Dr Hawkersworth who I Doubt not will do justice to ye work'.[15]

The 'Hawkersworth' referred to by Banks was Dr John Hawkesworth, whom the Admiralty commissioned to take the stilted, leaden journal

12. Cook's Journal, 13 July 1771.
13. Ibid.
14. Ibid.
15. Letter to the Comte de Lauraguais, December 1771, in Banks, *The Endeavour Journal of Joseph Banks*, Vol. 2, pp. 328-29; G. Williams, 'The Pacific: Exploration and Exploitation', in P.J. Marshall (ed.), *The Oxford History of the British Empire: Volume 2: The Eighteenth Century* (Oxford: Oxford University Press, 1998), p. 560.

entries of Cook and others and convert them into a stream of silvery prose. Hawkesworth had experienced some moderate success as an author and playwright by this time and through a mutual acquaintance – the composer and historian, Dr Charles Burney – was introduced to John Montagu, the Earl of Sandwich and First Lord of the Admiralty. Sandwich had previously mentioned to Burney that he had 'the papers of it [Cook's expedition] in his possession ... & ... that they were not arranged, but meer rough Draughts, & that he should be much obliged to any one who could recommend a proper Person to Write the Voyage'.[16] Burney had now found that 'proper Person'.

At the conclusion of their meeting, Hawkesworth was offered the job of editing the journals of various recent British expeditions (of which the *Endeavour* voyage would be the undoubted triumph), and was also given the rights to the material, meaning that he would be able to profit from their sale. In addition, he was handed £6,000 by the Admiralty for the work involved (a sum equivalent to a present value of over a million pounds).

Hawkesworth set to work on the base materials of Cook's expedition in India House, where he took an office to carry out the considerable task of preparing a three-volume publication from them. (A few years later, the essayist Charles Lamb also used the building as a place to write, but regarded it as a warren of passages, corridors and 'light-excluding, pent-up offices, where candles for one half the year supplied the place of the sun's light' and where shelves everywhere seemed piled up with manuscripts and files, which were unused and unremembered.[17]) These were the conditions in which Hawkesworth laboured away at Cook's records, sculpting them into a more publishable form for a general readership.

In June 1773, nearly two years after the *Endeavour* had returned to England, Hawkesworth's great work – encompassing the triumphs of British exploration and discovery of the age – was published.[18] Demand was keen but the critics' reaction was damning. Throughout the summer

---

16. F. Burney, *The Early Journals and Letters of Fanny Burney: Volume 1: 1768-1773*, ed. L.E. Troide (Kingston, ON: McGill-Queen's University Press, 1988), p. 173, in H.E. Parsons, 'Collaborative drawing on Captain Cook's *Endeavour* voyage, 1768-1771: An intellectual history of artistic practice' (PhD thesis, University of Melbourne, 2018), p. 67.
17. C. Lamb, 'The Superannuated Man', in C. Lamb, *Essays of Elia* (London: Macmillan & Co., 1883), p. 264.
18. J.L. Abbott, 'John Hawkesworth: Friend of Samuel Johnson and Editor of Captain Cook's Voyages and of *The Gentleman's Magazine*', *Eighteenth-Century Studies* 3, no. 3 (1970), p. 341.

of 1773, his work was attacked privately and publicly variously for its lack of interesting subject matter, for being overpriced, for making various technical errors, especially on nautical issues, for its diminution of some of the accomplishments of the voyages, and for adopting a style that was too elegant and therefore too estranged from the sort of language sea captains would use.[19] However, the sound and fury of its critics belies the fact that Hawkesworth's work continued to prove popular (running into several editions and initially selling over a thousand copies a month)[20] and remained the flagship work on the *Endeavour*'s expedition for more than a century.

The first volume of Hawkesworth's *An Account of the Voyages Undertaken by the Order of His Present Majesty for Making Discoveries in the Southern Hemisphere* dealt with the voyages of John Byron, Samuel Wallis and Philip Carteret, while the remaining two volumes were dedicated to the *Endeavour* expedition, based on the records of Cook and Banks[21] (although in order of work, Hawkesworth completed the *Endeavour* volumes first[22]). Cook departed for another voyage of discovery a month after Hawkesworth's book was published. He claimed that he 'never had the perusal of the Manuscript nor did I ever hear the whole of it read in the mode it was written'. However, there is an account by James Boswell of a conversation he had with Cook which suggests that he and Banks had at least some familiarity with the book as it was being prepared.[23] Nevertheless, regardless of the extent of Cook's knowledge of the Hawkesworth volumes, their publication represented the moment when the public got a first accurate glimpse of the shape of New Zealand (versions in French and German appeared the following year, ensuring a much larger readership for the work).[24]

19. Ibid., p. 342.
20. R.L. Ravneberg, 'The Hawkesworth Copy: An Investigation into the Printer's Copy Used for the Preparation of the 1773 Second Edition of John Hawkesworth's Account of Captain Cook's First Voyage', *Bibliographical Society of Australia and New Zealand Bulletin* 26, nos. 3/4 (2002), p. 173.
21. J. Hawkesworth, *An Account of the Voyages Undertaken by the Order of His Present Majesty for Making Discoveries in the Southern Hemisphere*, 3 vols (London: W. Strahan & T. Cadell, 1773).
22. W.H. Pearson, 'Hawkesworth's Alterations', *Journal of Pacific History* 7, no. 1 (1972), p. 49.
23. Cited in ibid., p. 57. This meeting is outlined in C.E. Vulliamy, *James Boswell* (London: Geoffrey Bles, 1932), p. 259.
24. J.O. Oliphant, 'A Project for a Christian Mission on the Northwest Coast of America, 1798', *Pacific Northwest Quarterly* 36, no. 2 (1945), p. 101. Oliphant also gives the details of the French and German translations.

For such an expensively produced and costly book to purchase (equivalent now to around £300) the quantity of maps included was disappointing (a point that Dalrymple made at the time).[25] For the volumes dealing with New Zealand, there were four charts of portions of the coast, seemingly selected at random, together with a map of New Zealand as a whole. This parsimony may have been enforced on Hawkesworth, though, as a consequence of not having been given access to all the maps and charts produced on the voyage. This, in turn, stems from the fact that the Admiralty did not have in its possession the complete set of maps made during the expedition. It is probable that most of the cartography produced by Cook was handed over to government officials or the Royal Society, but, as for the maps and journals of Banks, Pickersgill, Molyneux, Monkhouse and possibly others,[26] these ended up in private hands and their whereabouts eventually drifted into the realm of the unknown.[27] Although the Admiralty had been responsible for organising much of the finance for the expedition, it imposed no prohibition on officers making the information they collected available to private cartographers. At this time, there was still no system by which the Admiralty published the maps for which it, effectively, had paid. So, awkwardly, when the role of Hydrographer of the Navy was created in August 1795, the Admiralty found itself scraping together funds to buy much of the cartographical material which by right it should have owned in the first place.[28]

The four small charts of sections of New Zealand's coast that featured in the second volume of Hawkesworth's book were of Thames and Mercury Bay, the Bay of Islands, Tolaga Bay (all three of which appeared on one sheet in the book) and Cook Strait.[29] These contained the coordinates of their locations, a scale in leagues or miles, the depths of the waters along parts of the coast, the names given by Cook to various places, a smattering of topographical detail, concealed rocks off the coast and the route that the *Endeavour* took. These were all engraved

25. A. Dalrymple, in David (ed.), *The Charts and Coastal Views of Captain Cook's Voyages*, p. xlviii.
26. Such as C. Praval, 'Map of New Zeland [*sic*]', British Library, BL Add MS 7085 f. 1, record no. 10308.
27. J. Ross, *This Stern Coast: The Story of the Charting of the New Zealand Coast* (Wellington: A.H. & A.W. Reed, 1969), p. 49.
28. D.W. Haslam, 'Changing the Admiralty Chart', *The Journal of Navigation* 32, no. 2 (1979), p. 165; R. Morris, '200 Years of Admiralty Charts and Surveys', *The Mariner's Mirror* 82, no. 4 (1996), 420-435.
29. Hawkesworth, *An Account of the Voyages*, Vol. 2, pp. 322-24, 376.

by John Ryland – a skilled craftsman who rendered the images in very clear detail.

The iconic work in this volume was, of course, the map depicting the entirety of New Zealand. This territory was one of the last significant landmasses on earth to be charted and, until its publication by Hawkesworth, few outside the circle of those responsible for creating the map were aware of New Zealand's appearance and dimensions. Now, mystery had been displaced by certainty. The map that Cook had so expertly drawn was precisely replicated for printing by the engraver and cartographer, John Bayly.[30] This chart contained a more dramatic (although still largely figurative) representation of the country's topography and also displayed the *Endeavour*'s route, the depth of the water at selected locations (measured in fathoms), the full complement of names bestowed on places (with the exception of the North and South Islands, which still retained their Māori names at this stage), locations where the ship anchored, various coordinates, rocks above and just below the water line and, importantly, those few parts of the coast that Cook did not have time to explore and map.

Of course, for the public, this was much more than merely a geographical outline. The sparse details of the country's hinterland enabled New Zealand's shape to be populated by the imaginations of readers. It could be a South Pacific Arcadia peopled by noble savages, or an antipodean place of savage barbarity. There may not have been anything especially astounding about the country's fauna or flora, but its human geography was undoubtedly intriguing – certainly intriguing enough to the thousands of those wealthy enough to afford a copy of *An Account of ... Discoveries in the Southern Hemisphere*. To these readers, New Zealand was a major new territorial discovery roughly the same size as Britain, that was now partially known, yet still unclaimed by Europe. What could be gleaned from the maps and text was that the country was blessed with an abundance of harbours, a temperate climate, substantial stretches of fertile soil and dispersed throughout with communities living in 'villages', with their occupants proficient in a variety of traditional industries. For the more sanguine reader, New Zealand almost appeared to be in a state of rustic antiquity, just waiting its turn to be anointed with an infusion of British civilisation.

However, just how little political importance the Admiralty placed on New Zealand overall can be gauged from the fact that it did not get

30. Ibid., p. 280; B. Hooker, 'Official General Charts of New Zealand 1772-1885', *The Journal of Navigation* 41, no. 1 (1988), p. 35.

around to publishing its own version of Cook's chart of the country (based on the Bayly engraving) until 1816.[31] Previously, the Admiralty's only published map of part of New Zealand (of Doubtless Bay) was officially issued in 1781 but, inexplicably, based on a chart made by de Surville, rather than the superior charts by Cook that it had in its possession.[32] This signalled an almost dismissive approach both to New Zealand's significance to Britain at this time,[33] as well as to Cook's detailed mapping of the country, and, to an extent, this was as it should have been. The Admiralty's primary responsibility was the defence of the realm, which left little room for feeding popular curiosity in the region. After all, Hawkesworth's volumes had amply addressed that interest. Cook's careful cartographical work in New Zealand had been the fulfilment of the instructions issued to him and was not seen as the prologue for anything more.

31. Ross, *This Stern Coast*, p. 50; *A chart of New Zealand explored by Captain James Cook in 1769 and 1770 in His Majesty's Bark Endeavour* (1816), Auckland Libraries Heritage Collections Map 4979.
32. *Plan of the Bay of Lauriston, on New Zealand in 34°58'S: From a French M.S., December 1769* (London: A. Dalrymple [for the Admiralty Sept 20th], 1781), National Library of New Zealand, Alexander Turnbull Library, Neg. 146966 1/2 and Neg. C1402. The Bay of Lauriston is what Cook called Doubtless Bay; the plan drew on de Surville's voyage in the *Saint Jean-Baptiste*.
33. E.R. Bennett, 'Fidelity and Zeal: The Earl of Sandwich, Naval Intelligence, and the Salvation of Britain, 1763-1779', *The Historian* 70, no. 4 (2008), p. 672.

Chapter 12

Conclusion

For more than a century, the image of Tasman's sinuous line of New Zealand's west coast had radiated initially from printers' workshops in Amsterdam, gradually occupying maps and globes across Europe. The territory's presence twisted and turned in print as far as its form was concerned and was the subject of occasionally uninhibited speculation. Cook's map of New Zealand, published in London – by then Europe's new centre of mapmaking – straightened out the record. Cook had achieved the closest that a human could come to the act of Creation: bringing a landmass into being from a cartographical void.

For the remainder of the eighteenth century, a handful of European vessels made their way to New Zealand, but normally as part of some other larger expedition, rather than exclusively to explore the country. These tended more to be opportunistic excursions, though, usually undertaken by mariners of moderate abilities, following in the wake of masters such as Cook. Most of the vessels had at least one person on board who had some skills in cartography and, in the decades that followed, the final pieces of the mosaic of New Zealand's map were put in place.

At the same time, publishers throughout Europe were busy reconfiguring their maps and globes to include the completed image of New Zealand by resorting to the tried and tested technique of copying existing images with varying degrees of attention to detail (depending on the particular version of the map they had at hand). What was most likely the first reproduction of Cook's map of New Zealand (just preceding the publication of Hawkesworth's book) was a chart depicting the territories of the South Pacific. Like the Hawkesworth map, this too was engraved by Bayly but was specially commissioned by Banks, who paid for 100 copies of it to be printed: a few for his personal collection and the remainder for some of his friends and associates – a graphic representation (and perhaps boast) of the momentous expedition of which he had been a part.[1]

The French and German translations of Hawkesworth's work ensured that the march of knowledge about New Zealand proceeded at a steady pace throughout Europe. New copperplate engravings of the maps and charts of New Zealand based on those in the English edition of the book were produced and translated into these respective languages. In the French edition, published in 1774, the quality of the engravings (undertaken by Robert Bénard), although not up to Bayly's exquisite standard, was still very fine and contained all the location names, depth soundings and the *Endeavour*'s route around New Zealand that appeared on the English original.[2] The most immediately noticeable difference, though, was that in some of the French versions of this map, the country was coloured, which gave the images and the books housing them more commercial appeal.

A still more flamboyant image of the country appeared in the 1776 Italian map, *Nuove scoperte fatte nel 1765, 67 e 69 nel Mare del Sud*, which was published by the Venetian-based cartographer, Antonio Zatta (with the engraving done by Giuliono Zuliani).[3] The detail of New Zealand's coast in this work was poorly reconstructed and the place-names given by Cook to some of the locations were slightly misplaced, but, not only was this image more decorative and colourful than any other version up until this time, it was also the first mass-produced map to show New Zealand's location clearly in the context of the South Pacific.

The Italian connection extended to a visit to Fiordland in 1793 by two Spanish ships led by the Italian commander Alessandro Malaspina. This was a case of a map launching a journey, with Malaspina so inspired by Cook's accomplishments that he even named the two ships on his planned voyage after those used on Cook's expeditions – the *Descubierta*

1. Invoice received by Banks from John Bayly, for paper and engraving plate of map titled *The Great Pacific Ocean* (26 March 1772), Banks Papers (Series 06.057), State Library of New South Wales; J. Bayly, *The Great Pacific Ocean* (1772), British Library, Cartographic Items Maps 181.m.1, UIN: BLL01004959191.
2. J. Hawkesworth, *Relation des voyages entrepris par ordre de sa Majesté Britannique, actuellement regnante; Pour faire des découvertes dans l'hémisphère méridional*, Tome III (Paris: Saillant et Nyon, 1774), pl. 7.
3. Antonio Zatta, *Nuove scoperte fatte nel 1765, 67 e 69 nel Mare del Sud*, State Library of New South Wales, DDC M2 910/1770/1.

(*Endeavour*) and *Atevida* (*Resolution*). In February 1793, Malaspina ordered a boat to be sent from the *Descubierta* into Doubtful Sound with the expedition's cartographer Felipe Bauzá on board to produce a chart of the area. Bauzá fulfilled his instructions but, as had been the case previously with the British and the Dutch before them, the chart he created was filed away with all the other documents of this distant location when the expedition returned to Europe, and could have lain forgotten for generations. However, four years later, Bauzá was promoted to the position of Sub-Director of the Dirección de Hidrografía in Madrid. He promptly began to exhume many of the maps and documents that had been stored away, with the intention of producing an atlas that detailed Spanish territories and areas of interest in various parts of the world. It was a sort of imperial stock-take which was still underway in 1823, when a deterioration in the political situation in Spain forced Bauzá to seek refuge in England. The Admiralty was quick to recognise this 'asset' that had practically arrived in its lap and immediately sent him to work in the Hydrographic Office. Among the abundance of intelligence that Bauzá syphoned from Spain to Britain was his chart of Doubtful Sound, which officially entered the archives of the Admiralty in 1830 and, from that point, made its way into British maps of New Zealand.[4]

Cook's charts of New Zealand had been so thorough that, in the remaining years of the eighteenth century, subsequent expeditions which visited the territory only added small quantities of additional cartographical detail. The final British voyage to New Zealand in the eighteenth century was led by Captain George Vancouver, who had previously sailed with Cook on his 1772-75 circumnavigation of the world. In 1791, Vancouver explored Dusky Sound (which lay around 60 kilometres south of Doubtful Sound), along with Lieutenant William Broughton, who was in command of the other ship on this expedition. The two prodded the harbours and inlets, taking measurements, delineating the form of the coast in the sound, and attaching names to various places in the process. In due course, this small mapping exercise was published by the Admiralty and another detail of a previously cursory outline of the country was slotted into place,[5] although it took

4. R.J. King, 'Puerto del Pendulo, Doubtful Sound: The Malaspina Expedition's Visit to New Zealand in Quest of the True Figure of the Earth', *The Globe* 65, no. 1 (2010), p. 220.
5. W.R. Broughton, *Dusky Bay in New Zealand. As Copied from a correct Sketch of Captain Cook's, with the termination of two Arms from Apparent Isles, examined by the Discovery & Chatham's Boats in 1791 by Lieut. Broughton,*

the Admiralty until 1840 to make the necessary changes to its charts and publish them.[6]

Then there was the cartography that never made it into the finalised maps of New Zealand published in the eighteenth century. The most notable example of this was the work carried out on the expedition led by Marc-Joseph Marion du Fresne, who visited northern New Zealand in 1772. One of the commanders on this voyage, Ambroise Bernard Marie le Jar du Clesmeur, produced at least two charts of parts of the North Island,[7] including one of the Bay of Islands, which was titled 'Plan du Port Marion', based on the French name bestowed on the location (and where du Fresne and 24 crew members of the expedition were killed in an encounter with local Māori in June 1772).[8] Du Clesmeur's map of the Bay of Islands[9] was more detailed than Cook's but, shortly after it was printed in Paris, possibly in the early 1780s,[10] it disappeared from circulation. Such losses were not uncommon in this era.[11] Given the (probably) small print-run, and the obscure subject-matter, the interest in preserving such items often died out when their collectors did. The hand-coloured version of this particular chart only resurfaced in the public domain in 1950, when it was presented to the Peabody Museum at Yale University by an anonymous donor.[12] The absence of

---

National Library of New Zealand, ref. MapColl-NZGB-4/23/309/Acc.54995; G. Vancouver, *A Voyage of Discovery to the North Pacific Ocean, and Round the World ... Performed in the Years 1790, 1791, 1792, 1793, 1794 and 1795* (London: John Stockdale, 1801), pp. 183-86.

6. Ross, *This Stern Coast*, p. 61.
7. A.B.M. du Clesmeur, 'Sketch of Mt Taranaki' (28 March 1772), in Lee, Navigators & Naturalists, p. 147.
8. McNab, *Historical Records of New Zealand*, Vol. 2, pp. 413-21.
9. A.B.M. du Clesmeur, 'Plan du Port Marion à la Nouvelle Zelande' (1772?), National Library of New Zealand, ref MapColl-832.11aj/[1772]/Acc.814.
10. The estimation of this period is based on the publication in Paris of 'Plan du Port Marion à la Nouvelle Zelande', [ms map], and 'Plan des terres découvertes par M. de Surville qu'il a nommé Côte des Assacides', [ms map]. These are two ink sketches which were probably preparatory to engravings for the corresponding maps published in 1783 in J.-M. Crozet, *Nouveau voyage à la mer du Sud commencé sous les ordres de M. Marion* (Paris: Barrois, 1783), p. 293. The two ink sketches are in National Library of New Zealand, ref. MapColl-832.11aj/1772/Acc.31500-1.
11. E.M. Hallam, 'Problems with Record Keeping in Early Eighteenth-Century London: Some Pictorial Representations of the State Paper Office, 1705-1706', *Journal of the Society of Archivists* 6, no. 4 (1979), 219-26.
12. Lee, *Navigators & Naturalists*, p. 225.

this work from widespread circulation in the final decades of the eighteenth century obviously meant that it was out of reach to the British Admiralty and, consequently, was unable to inform their mapmaking in this period.

Incidental visits to New Zealand around this time, such as Antoine d'Entrecasteaux's 1791 voyage which edged along the northern tip of the country, resulted in a survey of the coast from Cape Maria van Diemen to the Surville Cliffs being completed by the expedition's cartographer, Charles-François Beautemps-Beaupré.[13] However, this area had already been charted by Cook and had been represented to the satisfaction of the Admiralty in Hawkesworth's book eight years earlier. Consequently, Beautemps-Beaupré's chart, like du Clesmeur's before him, ended up being drafted to blush unseen by most people. It was the Cook map of New Zealand which became the template for others to emulate in the decades that followed its publication.

* * *

Throughout the following century, the map of New Zealand continued to be added to and corrected. However, it was with the publication of Cook's map in 1773 that its form, which had remained fluid for around 130 years, was finally fixed. Subsequent cartographers refined the work, but never superseded the magnitude of its accomplishment – something which Cook's pedantic attention to the smallest of details ensured. Moreover, all this intelligence – gathered at great cost – was immediately disclosed fully and publicly. The image of New Zealand went from cartographical blankness in the sixteenth century, to a suggestive though slightly incoherent presence in the seventeenth, to full visibility in the eighteenth. This act of outlining a territory also brought into view what would be looked on covetously by some Europeans in the nineteenth century. How could it be any other way? By the early 1800s many Britons, in particular, regarded the world as some sort of vast commercial emporium.[14]

---

13. H. Richard, 'The Interest of French Cartography in the South Pacific in the Eighteenth Century', *The Globe,* 83 (2018), 1-11.
14. S. Raffles to T. Addenbrooke, Singapore, 10 June 1819, in T.S. Raffles, 'The Founding of Singapore', *Journal of the Straits Branch of the Royal Asiatic Society* 2 (December 1878), p. 175; P. Moon, 'Entering the Periphery: Reassessing British Involvement in New Zealand in the 1820s in the Context of Wallerstein's Theory of a World System', *New Zealand Journal of History* 49, no. 2 (2015), 81-109; Wallerstein, 'The Rise and Future Demise of the World Capitalist System', 387-415.

Consequently, sketching the locations and dimensions of territories in different parts of the world amounted to an act of ideological geography: stretching the horizon of imperial prospects.

* * *

All this was still far into the future, though, when Cook's map of New Zealand first reached its mass market in 1773. Its appearance unavoidably piqued popular curiosity – there is always magic in mystery – and the satiation of that curiosity could only be achieved through further exploration. To that extent, the first full map of New Zealand, created in the narrow cabins of the *Endeavour* in 1769 and 1770, both completed the task of defining the country's shape and ensured that further European probing and penetration was almost certain. The long history of the emergence of the country's first map, which ended with Cook's chart, would end up being just the prelude to that process.

Bibliography

Maps and Images

A chart of New Zealand explored by Captain James Cook in 1769 and 1770 in His Majesty's Bark Endeavour (1816), Auckland Libraries Heritage Collections Map 4979

'Chart of the Society Islands' (possibly drawn by Joshua Banks in collaboration with Tupaia), British Library, BL Add MS 15508, no. 18

Orbis terrarum typus de integro in plurimis emendatus, auctus, et icunculis illustratus, Sanders of Oxford, Antique Prints and Maps, stock I.D. 43111

Plan du Port Marion à la Nouvelle Zelande [ms map] and *Plan des Terres découvertes par Mr. de Surville qu'il a nommé Côtes des Assacides* [ms map]; these are two ink sketches which were probably preparatory to engravings for the corresponding maps published in 1783 in J.-M. Crozet, *Nouveau voyage à la mer du Sud commencé sous les ordres de M. Marion* (Paris: Barrois, 1783); both these maps are also in the collection of the National Library of New Zealand, ref. MapColl-832.11aj/1772/Acc.31500-1; *Plan du Port Marion* (now known as the Bay of Islands) is attributed by the National Library to A.B.M. du Clesmeur.

Plan of the Bay of Lauriston, on New Zealand in 34°58'S: from a French M.S., December 1769 (London: A. Dalrymple [for the Admiralty Sept 20th], 1781), National Library of New Zealand, Alexander Turnbull Library Neg. 146966 1/2 and Neg. C1402

Soli Britannico Reduci Carolo Secundo regum augustissimo hoc Orbis Terrae Compendium humill. off. I. Klencke (A collection of forty-two maps of all parts of the world, published by J. and W. Blaeu, H. Allard, N.J. Visscher and others, made up into a volume by J. Klencke and other merchants of Amsterdam and presented by them to King Charles II of England at his accession in 1660), 1613-1660, British Library, System number 004959010, UIN: BLL01004959010

Bayly, J., *The Great Pacific Ocean* (1772), British Library, Cartographic Items Maps 181.m.1, UIN: BLL01004959191

Beeckman, A., *The Castle of Batavia, seen from Kali Besar West*, Rijksmuseum, Object No. SK-A-19

Bellin, J.N., *Carte réduite des Terres Australes* (Paris, 1753), Stanford Libraries, Barry Lawrence Ruderman Map Collection

Blaeu, J., *Archipelagus Orientalis sive Asiaticus* (Amsterdam: Joan Blaeu, 1659), paper on linen with wooden rollers; large wall map of Southeast Asia and Australia in six engraved map-sheets, three panels of letterpress descriptive text in Latin, Dutch and French with imprint of Blaeu in the three languages, inset maps of Borin and the Solomon islands, contemporary hand colour, backed on original linen, original decorated rollers, total dimensions 158.7 × 117.4cm, Sotheby's lot 93 (May 2017)

Blaeu, J., *Nova totius terrarum orbis tabula* (1648), Harry Ransom Center, University of Texas at Austin, Map_Kraus_03

Blaeu, W., and J. Blaeu, *Nova et accurata totius terrarum orbis tabula* (1645-46 [1619]), Collecties Maritiem Museum 'Prins Hendrik', Rotterdam, ref. K259

Bowen, E., *A Complete Map of the Southern Continent Survey'd by Capt. Abel Tasman & depicted by Order of the East India Company in Holland in the Stadt House at Amsterdam* (London, 1744), Stanford Libraries, Barry Lawrence Ruderman Map Collection

Broughton, W.R., *Dusky Bay in New Zealand. As Copied from a correct Sketch of Captain Cook's, with the termination of two Arms from Apparent Isles, examined by the Discovery & Chatham's Boats in 1791 by Lieut. Broughton*, National Library of New Zealand, ref. MapColl-NZGB-4/23/309/Acc.54995

Callander, J., *Reduced Chart of Australasia*, in J. Callander and C. de Brosses, *Terra Australis Cognita, or, Voyages to the Terra Australis, or Southern Hemisphere, During the Sixteenth, Seventeenth, and Eighteenth Centuries* (Edinburgh: J. Donaldson, 1768)

Clesmeur, A.B.M. du, *Plan du Port Marion*, A.B.M. du Clesmeur, 'Plan du Port Marion à la Nouvelle Zelande' (1772?), National Library of New Zealand, ref MapColl-832.11aj/[1772]/Acc.

Clesmeur, A.B.M. du, *Sketch of Mt Taranaki* (28 March 1772), AN Serie Marine B4-317 Pièce 61, 52, Archives Nationales Paris

Cuyck, J.C. van, *Batavia in 1629 tijdens de Mataramse belegering*, General State Archives of The Hague, Maps and Drawings Department, collection VEL, inv.nr. 1179 A

Cuyp, J.G., *Portrait of Abel Tasman, his wife and daughter* (*c.* 1637), National Library of Australia, call number PIC T267 NK3

Gerritsz, H., *Extrait de la carte du Pacifique d'Hessel Gerritsz (1622), montrant les 'îles Salomon' ainsi que 5 îles découvertes par Jacob Le Maire et Willem Schouten en 1616*, Bibliothèque nationale de France, Carte conservée au département des Cartes et plans, cote GE SH ARCH-30 (RES)

Gerritsz, H., *Mar del Sur. Mar Pacifico*, Bibliothèque nationale de France, Département des Cartes et plans, cote GE SH ARCH-30

Gerritsz, H., *Plattegrond van Batavia en omstreken*, National Archives of the Netherlands, NL-HaNA_4.VEL_1179B

Gilsemans, I., *A view of the* Murderers' *Bay, as* you *are at* anchor *here in 15 fathom* (1642), National Library of New Zealand, Alexander Turnbull Library, ref. PUBL-0086-021

Goos, P., *Pascaerte Vande Zuyd-Zee tussche California, en Ilhas de Ladrones* (1666), Yale University Library, Beinecke Rare Book and Manuscript Library, Call Number 23cea 1666, Container/Volume BRBL_00682

Hondius, H., *Polus Antarcticus* (Amsterdam, 1637), State Library of Western Australia, Globe, no. 54, 2003 0311-3930, CIU file ref. 16/212

Janszoon, J., *Terra Australis Incognita* (1652), Harvard Library, Cambridge MA, Harvard Map Collection, ref. MAP-LC G9801.S12 1652. J3

Keulen, J. van, *Pascaert vande Zuyd Zee en een gedeelte van Brasil van ilhas de Ladronos tot R. de la Plata* (1680), Yale University Library, Beinecke Rare Book and Manuscript Library, Call Number 23cea 1697, Container/Volume BRBL_00682

Moxon, J., *A plat of all the world: projected according to the truest rules being far more exact than either the plain-card or the maps of the world described in two rounds*, Boston Public Library, Mapping Boston Collection, Identifier: 06_01_000090, Barcode: 39999052509518

Parkinson, S., *A View of the North Side of the Entrance into Poverty Bay & Morai Island in New Zealand. 1. Young Nick's Head. 2. Morai Island. View of another Side of the Entrance into the said Bay*, S. Parkinson del. R.B. Godfrey sc. Plate XIV [London, 1784], National Library of New Zealand, ref. PUBL-0037-14

Pickersgill, R., *A chart of part of the So. Contit. between Poverty Bay and the Court of Aldermen discovered by His Maj.s Bark Endeavour* [copy of ms map] [1769], National Library of New Zealand, ref. MapColl-832.1aj/[1769]/Acc.12471

Pickersgill, R., *New Zealand: Capt. Cook's Discoveries* [1771], National Library of Australia, Charts of the Hydrographic Department (as filmed by the AJCP) [microform] : pre-1825 :[M406], 1770-1824./File 139-40.458

Praval, C., *Map of New Zeland* [*sic*], British Library, BL Add MS 7085 f. 1, record no. 10308

Prinald, *A New Map of the World on Mercator's Projection* (1766), National Library of Australia, Bib ID, 1773664

Seale, R., *A Chart Shewing the Track of the* Centurion *Round the World* (London, *c*. 1744), National Library of Australia, Rex Nan Kivell Map Collection, Call number MAP NK 4592

Seller, J., *A Chart of the South-Sea* (London, *c*. 1672), Boston Public Library, Norman B. Leventhal Map Center, Identifier: 06_01_008267, Call #:G1059 .S45 1672

Spöring, H., *A Fortified Town or Village called a Hippah, built on a Perforated Rock, at Tolaga in New Zealand,* Morris sculp. (London. Alexr. Hogg [1784?]), National Library of New Zealand, ref. B-098-025

State Library of New South Wales, *The Tasman Map* (Sydney: State Library of New South Wales, 2020)

Tasman, A., *Australia and New Zealand [cartographic material]: from the original map made under the direction of Abel Tasman in 1644 and now in the Mitchell Library, Sydney / this facsimile was drawn by James Emery in 1946* (Sydney: Trustees of the Public Library of New South Wales, 1946), ref. rbr293624

Tasman's kaart van zijn Australische ontdekkingen 1644, *de Bonaparte-kaart,* gereproduceerd op de ware grootte in goud en kleuren naar het origineel in de Mitchell Library, Sydney; met toestemming van de autoriteiten door F.C. Wieder, State Library of New South Wales, ref. MAP NK 1791

Tupaia's Map, 1770, British Library, BL Add MS 21593.C (T3/B)

Vaugondy, R. de, *Carte réduite de l'Australasie, pour servir à la lecture de l'histoire des terres Australes [cartographic material] / par le Sr. Robert de Vaugondy, Geog. ord. du Roi, de l'Academie Royale des Sciences et Belles-Lettres, de Nancy, 1756; [engraved by] G. de-la-Haye,* National Library of Australia, Bib ID. 1190483

Vennekool, J., *De Grondt en Vloer vande Groote Burger Sael; Tot Amsterdam: Bij Dancker Danckerts* (1661), State Library of New South Wales, M2 100/1661/1

Wit, F. de, *Nova totius terrarum orbis tabula* (1660), Library of Congress Geography and Map Division, Washington, DC, G3200 1660.W5, Library of Congress Control Number 2006627253

Zatta, A., *Nuove scoperte fatte nel 1765, 67 e 69 nel Mare del Sud,* State Library of New South Wales, DDC M2 910/1770/1

Unpublished Material

Algemeen Rijksarchief, Koloniaale Afdeling (State Archives, Colonial Department, The Hague), 452, Missive Heren XVII to J.Pz. Coen, 9 September 1620; 4653 D 207 b. Missive by Hessel Gerritsz to Adriaen van Santvoort, 23 June 1622

Bowrey, T., Correspondence, diaries, drawings, charts, maps and other papers, London Metropolitan Archives, CLC/427/MS24176 (microfilm MS24177/001)

Charters of the East India Company with related documents: the 'parchment records', British Library, Asian and African Studies, ref. IOR/A/1

Commissioners for executing the office of Lord High Admiral of Great Britain, 'Secret. Additional Instructions for Lt James Cook, Appointed to Command His Majesty's Bark the Endeavour', MS 2-Cook's voyage 1768-71 [manuscript]: copies of correspondence, etc., National Library of Australia, Bib ID: 1120886

Cook, J., 'Revised holograph manuscript journal, *HMS Resolution* (1775)', British Library, Department of Manuscripts, BL Add MS 27888
Copy of the journal of a voyage from Batavia, in the East Indies, for the discovery of the unknown South land, by Abel Jansen Tasman, with sketches of the coast and peoples, British Library, BL Add MS 8946
Invoice received by Banks from John Bayly, for paper and engraving plate of map titled *The Great Pacific Ocean* (26 March 1772), Banks Papers (Series 06.057), State Library of New South Wales
'Journaal van Abel Jansz. Tasman van zijn ontdekkingsreis naar het Zuidland, 1642-1643', Zie Aanwinsten, 1.11.01.01, inv. nr. 121 (1867 A III); 'Tekeningen behorende bij het journaal van Abel Jansz. Tasman van de ontdekking van Van Diemensland en Nieuw-Zeeland', Zie Aanwinsten, 1.11.01.01, inv. nr. 320 (1886 B XV), Nationaal Archief, The Hague.
Letter from Bantam, 10 January 1644, British Library, India Office Records, series IOR/L/MAR/A
Royal Society, The, Repository GB 117, ref. CMO/1/46 (30 March 1664)
Tasman, A., 'Extract Uittet Journael vanden Scpr Commandr Abel janssen Tasman, bij hem selffs int ontdecken van't onbekende Zuijdlandt gehouden' (1642-43), State Library of New South Wales, Safe 1/72, ref. 423571
Wallis, S., 'Log books and sketchbook kept during his voyage around the world in command of H.M.S. Dolphin', British Admiralty (Adm.#55/35, PRO Reel 1579), in National Library of New Zealand, Micro-MS-0353

Miscellaneous

Greenhill, B., 'Captain James Cook, RN', Address at Cook Commemorative Service, Westminster Abbey, 11 February 1979
Twelve-inch quadrant by John Bird (London, 1760-69), Science Museum Group, Object Number 1876-572/1

Theses

Ariese, C.E., 'A twisted truth – the VOC ship *Batavia*: Comparing history & archaeology' (BA dissertation, Leiden University, 2010)
Byloos, B., 'Nederlands vernuft in Spaanse dienst: Technologische bijdragen uit de Nederlanden voor het Spaanse Rijk, 1550-1700' (PhD thesis, Katholieke Universiteit, Leuven, 1986)
Cook, A.S., 'Alexander Dalrymple (1737-1808), hydrographer to the East India Company and the Admiralty, as publisher: a catalogue of books and charts (PhD thesis, University of St Andrews, 1993)
Lanegran, D.A., 'Alexander Dalrymple: Hydrographer (PhD thesis, University of Minnesota, 1970)
Luminet, J.-P., 'Willem Janszoon Blaeu' (unpublished paper, Laboratoire Univers et Théories, Paris, 2015)

Parsons, H.E., 'Collaborative drawing on Captain Cook's *Endeavour* voyage, 1768-1771: An intellectual history of artistic practice (PhD thesis, University of Melbourne, 2018)

Conference Proceedings

Beaglehole, J.C., 'Some problems of editing Cook's journals', paper read to the history section of the Australian and New Zealand Association for the Advancement of Science (Dunedin, January 1957)

Farrauto, L., and P. Ciuccarelli, 'The image of the divided city through maps: The territory without territory' (Indaco Department, Politecnico di Milano, January 2021)

Journal Articles and Reports

Abbott, J.L., 'John Hawkesworth: Friend of Samuel Johnson and Editor of Captain Cook's Voyages and of *The Gentleman's Magazine*', *Eighteenth-Century Studies* 3, no. 3 (1970)

Andrade, T., 'The Company's Chinese Pirates: How the Dutch East India Company Tried to Lead a Coalition of Pirates to War against China, 1621-1662', *Journal of World History* 15, no. 4 (2004), 415-44

Anonymous, 'Biographical Memoir of Alexander Dalrymple, Esq., Late Hydrographer to the Admiralty', *The Naval Chronicle* 35 (1816), 177-204

Anonymous, 'Memoirs of Alexander Dalrymple Esq.', *The European Magazine and London Review* 42 (November 1802), 323-28

Axtell, J., 'Europeans, Indians, and the Age of Discovery in American History Textbooks', *American Historical Review* 92, no. 3 (1987), 621-32

Baena, L.M., 'Negotiating Sovereignty: The Peace Treaty of Münster, 1648', *History of Political Thought* 28, no. 4 (2007), 617-41

Barritt, M.K., 'Navigational Enterprises in Europe and Its Empires, 1730-1850', *The Mariner's Mirror* 102, no. 2 (2016), 231-33

Bayly, C., 'Knowing the Country: Empire and Information in India', *Modern Asian Studies* 27, no. 1 (1993), 3-43

Beaglehole, J.C., 'Cook the Navigator' (A lecture delivered to the Royal Society on 3 June 1969 on the occasion of the celebration of the observation of the transit of Venus by Captain James Cook, R.N., F.R.S.', *Proceedings of the Royal Society of London (Series A, Mathematical and Physical Sciences)* 314, no. 1516 (1969), 27-38

Beasley, A.W., 'The First Amputation in Australia', *Seventeenth Century* 17, no. 1 (2002), 70-77

Bennett, E.R., 'Fidelity and Zeal: The Earl of Sandwich, Naval Intelligence, and the Salvation of Britain, 1763-1779', *The Historian* 70, no. 4 (2008), 669-96

Bennett, M., 'Van Diemen, Tasman and the Dutch Reconnaissance', *Papers and Proceedings: Tasmanian Historical Research Association* 39, no. 2 (1992)

Borch, M., 'Rethinking the Origins of Terra Nullius', *Australian Historical Studies* 32, no. 117 (2001), 222-39
Broecke, M.P.R. van den, 'Unstable Editions of Ortelius' Atlas', *Map Collector* 70 (1995), 2-8
Bruijn, J.R., 'Between Batavia and the Cape: Shipping Patterns of the Dutch East India Company', *Journal of Southeast Asian Studies* 11, no. 2 (1980), 251-65
Buringh, E., and J.L. van Zanden, 'Charting the "Rise of the West": Manuscripts and Printed Books in Europe, a Long-Term Perspective from the Sixth through Eighteenth Centuries', *The Journal of Economic History* 69, no. 2 (2009), 409-45
Campbell, T., 'A Descriptive Census of Willem Blaeu's Sixty-Eight Centimetre Globes', *Imago Mundi* 28, no. 1 (1976), 21-50
Chapman, A., 'The Accuracy of Angular Measuring Instruments Used in Astronomy between 1500 and 1850', *Journal for the History of Astronomy* 14, no. 2 (1983)
Charlier, R.H., and C. Charlier, 'Lowlands Sixteenth Century Cartography: Mercator's Birth Pentecentennial', *Journal of Coastal Research* 32, no. 3 (2016), 670-85
Clay, D., 'Plato's Atlantis: The Anatomy of a Fiction', *Proceedings of the Boston Area Colloquium in Ancient Philosophy* 15, no. 1 (1999), 1-21
Cock, R., 'Precursors of Cook: The Voyages of the *Dolphin*, 1764-8', *The Mariner's Mirror* 85, no. 1 (1999), 30-52
Collingridge, V., 'Mapping the Fantastic Great Southern Continent, 1760-1777: A Study in Enlightenment Geography', *The Cartographic Journal* 57, no. 4 (2020), 335-52
Conway, M., 'The Cartography of Spitsbergen', *Geographical Journal* 21, no. 6 (1903), 636-44
Cook, J., and C. Green, 'Observations Made, by Appointment of the Royal Society, at King George's Island in the South Sea; By Mr. Charles Green, Formerly Assistant at the Royal Observatory at Greenwich, and Lieut. James Cook, of His Majesty's Ship the Endeavour', *Philosophical Transactions (1683-1775)* 61 (1771), 397-421
Craib, R.B., 'Cartography and Power in the Conquest and Creation of New Spain', *Latin American Research Review* 35, no. 1 (2000), 7-36
Croarken, M., 'Providing Longitude for All', *Journal for Maritime Research* 4, no. 1 (2002), 106-26
Crone, G.R., 'The Discovery of Tasmania and New Zealand', *The Geographical Journal* 111, nos. 4/6 (1948), 257-63
Cross, E., 'The Last French East India Company in the Revolutionary Atlantic', *The William and Mary Quarterly* 77, no. 4 (2020)
Cross, R.W., 'Dutch Cartographers of the Seventeenth Century', *Geographical Review* 6, no. 1 (1918), 66-70
Davis, J., and C. Daniel, 'John Seller: Instrument Maker and Plagiarist', *Bulletin of the Scientific Instrument Society* 102 (2009)

Debergh, M., 'A Comparative Study of Two Dutch Paps, Preserved in the Tokyo National Museum: Joan Blaeu's Wall Map of the World in Two Hemispheres, 1648, and Its Revision *ca.* 1678 by N. Visscher', *Imago Mundi* 35, no. 1 (1983), 20-36

Diller, A., 'The Oldest Manuscripts of Ptolemaic Maps', *Transactions and Proceedings of the American Philological Association* 71 (1940), 62-67

Di Piazza, A., and E. Pearthree, 'A New Reading of Tupaia's Chart', *The Journal of the Polynesian Society* 116, no. 3 (2007), 321-40

Doe, N.A., 'The Potential Accuracy of the Eighteenth-Century Method of Determining Longitude at Sea', *SILT – A Journal of Personal Research* 556 (2016), 1-6

Douglas, B., 'Terra Australis to Oceania: Racial Geography in the "Fifth Part of the World"', *Journal of Pacific History* 45, no. 2 (2010), 179-210

Eckstein, L., and A. Schwarz, 'The Making of Tupaia's Map: A Story of the Extent and Mastery of Polynesian Navigation, Competing Systems of Wayfinding on James Cook's *Endeavour*, and the Invention of an Ingenious Cartographic System', *Journal of Pacific History* 54, no. 1 (2019), 1-95

Edney, M.H., 'Putting "Cartography" into the History of Cartography: Arthur H. Robinson, David Woodward, and the Creation of a Discipline', *Cartographic Perspectives* 51 (2005), 14-29

Edney, M.H., 'Mathematical Cosmography and the Social Ideology of British Cartography, 1780-1820', *Imago Mundi* 46, no. 1 (1994), 101-16

Emmer, P.C., 'The First Global War: The Dutch versus Iberia in Asia, Africa and the New World, 1590-1609', *Journal of Portuguese History* 1, no. 1 (2003)

Erskine, N., 'The *Endeavour* after James Cook: The Forgotten Years, 1771-1778', *The Great Circle* 39, no. 1 (2017), 55-88

Fitzmaurice, A., 'The Genealogy of Terra Nullius', *Australian Historical Studies* 38, no. 129 (2007), 1-15

Foster, W., 'An Early Chart of Tasmania', *Geographical Journal* 37, no. 5 (1911), 550-51

Fry, H.T., 'Alexander Dalrymple and New Guinea', *Journal of Pacific History* 4, no. 1 (1969), 83-104

Gaastra, F.S., 'The Dutch East India Company: A Reluctant Discoverer', *The Great Circle: Journal of the Australian Association for Maritime History* 19, no. 2 (1997), 109-23

Gascoigne, J., 'The Royal Society and the Emergence of Science as an Instrument of State Policy', *The British Journal for the History of Science* 32, no. 2 (1999), 171-84

Gawronski, J., 'East Indiaman *Amsterdam* Research 1984-1986', *Antiquity* 64, no. 243 (1990), 363-75

Gelderblom, O., and J. Jonker, 'Completing a Financial Revolution: The Finance of the Dutch East India Trade and the Rise of the Amsterdam Capital Market, 1595-1612', *The Journal of Economic History* 64, no. 3 (2004), 641-72

Gelderblom, O., A. de Jong and J. Jonker, 'The Formative Years of the Modern Corporation: The Dutch East India Company VOC, 1602-1623', *The Journal of Economic History* 73, no. 4 (2013), 105-76

Gentleman's Magazine, *The*, 10 (London, January 1740)
Gentleman's Magazine, *The*, 22 (London, August 1752)
Gentleman's Magazine, *The*, 33 (London, 1763)
Gerritsen, R., 'Getting the Strait Facts Straight', *The Globe* 72 (2013), 11-21
Gitzen, G.D., 'Edward Wright's World Chart of 1599', *Terrae Incognitae* 46, no. 1 (2014), 3-15
Glover, W., 'The Eighteenth Century Practice of Navigation as Recorded in the Logs of Hudson's Bay Company Ships', *The Northern Mariner/Le marin du nord* 26, no. 2 (2016), 145-64
Gomperts, A., A. Haag and P. Carey, 'Mapping Majapahit: Wardenaar's Archaeological Survey at Trowulan in 1815', *Indonesia* 93 (2012), 177-96
Graham, G., 'Observations of the Dipping Needle, Made at London, in the Beginning of the Year 1723. By Mr. George Graham, Watchmaker, FRS', *Philosophical Transactions (1683-1775)* 33 (1724), 332-39
Haarhoff, J., 'Water and Beverages on DEIC Ships between the Netherlands and the Cape: 1602-1795', *Historia* 52, no. 1 (2007), 127-54
Haft, A.J., 'Imagining Space and Time in Kenneth Slessor's "Dutch Seacoast" and Joan Blaeu's Town Atlas of The Netherlands: Maps and Mapping in Kenneth Slessor's Poetic Sequence *The Atlas*, Part Three', *Cartographic Perspectives* 74 (2013), 29-54
Hallam, E.M., 'Problems with Record Keeping in Early Eighteenth-Century London: Some Pictorial Representations of the State Paper Office, 1705-1706', *Journal of the Society of Archivists* 6, no. 4 (1979), 219-26
Hargrave, J., 'Joseph Moxon: A Re-fashioned Appraisal', *Script & Print* 39, no. 3 (2015), 163-81
Haslam, D.W., 'Changing the Admiralty Chart', *The Journal of Navigation* 32, no. 2 (1979), 164-70
Heawood, E., 'A Masterpiece of Joan Blaeu', *Geographical Journal* 55, no. 4 (1920), 312-15
Hejeebu, S., 'Contract Enforcement in the English East India Company', *The Journal of Economic History* 65, no. 2 (2005), 496-523
Henderson, F., 'Robert Hooke and the Visual World of the Early Royal Society', *Perspectives on Science* 27, no. 3 (2019), 395-434
Herdendorf, C.E., 'Captain James Cook and the Transits of Mercury and Venus', *Journal of Pacific History* 21, no. 1 (1986), 39-56
Hirschmann, J.V., and J.R. Gregory, 'Adult Scurvy', *Journal of the American Academy of Dermatology* 41, no. 6 (1999), 895-906
Hocken, T.M., 'Abel Tasman and His Journal', *Transactions and Proceedings of the Royal Society of New Zealand* 28 (1895)
Hooker, B., 'James Cook's Secret Search in 1769', *The Mariner's Mirror* 87, no. 3 (2001), 297-302
Hooker, B., 'New Light on Jodocus Hondius' Great World Mercator Map of 1598', *Geographical Journal* 159, no. 1 (1993), 45-50
Hooker, B., 'New Light on the Mapping and Naming of New Zealand', *New Zealand Journal of History* 6, no. 2 (1972), 158-67

Hooker, B., 'Official General Charts of New Zealand 1772-1885', *The Journal of Navigation* 41, no. 1 (1988), 35-51

Hooker, B., 'Towards the Identification of the Terrestrial Globe Carried on the *Heemskerck* by Abel Tasman in 1642-43', *The Globe* (Journal of the Australian and New Zealand Map Society) 79 (2016), 31-37

Hooker, B., 'Two Sets of Tasman Longitudes in Seventeenth and Eighteenth Century Maps', *Geographical Journal* 156, no. 1 (1990), 23-30

Hornsby, T., 'XXXIV. On the transit of Venus in 1769. To the Right Honourable The Earl of Morton, President to the Council and Fellows of the Royal Society, this discourse is, with all humility, inscribed, by their humble servant, Thomas Hornsby', *Philosophical Transactions* 55 (1765), 326-44

Houston, R., 'Literacy and Society in the West, 1500-1850', *Social History* 8, no. 3 (1983), 269-93

Huigen, S., 'Repackaging East Indies Natural History in François Valentyn's *Oud en Nieuw Oost-Indiën*', *Early Modern Low Countries* 3, no. 2 (2019), 234-64

Hunt, M., 'Racism, Imperialism, and the Traveler's Gaze in Eighteenth-Century England', *Journal of British Studies* 32, no. 4 (1993), 333-57

Hunter, M., 'The Crown, the Public and the New Science, 1689-1702', *Notes and Records of the Royal Society of London* 43 (1989), 99-116

Iliffe, R., 'Material Doubts: Hooke, Artisan Culture and the Exchange of Information in 1670s London', *British Journal for the History of Science* 28, no. 3 (1995), 285-318

Irwin, G., D. Johns, R. Flay *et al.*, 'A Review of Archaeological Maori Canoes (*Waka*) Reveals Changes in Sailing Technology and Maritime Communications in Aotearoa/New Zealand, AD 1300-1800', *Journal of Pacific Archaeology* 8, no. 2 (2017), 31-43

Isdale, L., 'Janszoon in Context: Cartography from Earliest Times to Flinders', *Queensland History Journal* 21, no. 2 (2010), 103-16

Jagger, G., 'Joseph Moxon, FRS, and the Royal Society', *Notes and Records of the Royal Society of London* 49, no. 2 (1995), 193-208

James, K., 'Reading Numbers in Early Modern England', *BSHM Bulletin* 26, no. 1 (2011)

Janzen, O.U., 'The Making of a Maritime Explorer: James Cook in Newfoundland, 1762-1767', *The Northern Mariner/Le marin du nord* 28, no. 1 (2018), 23-38

Kan, J. van, 'De Bataviasche statuten en de buitencomptoiren', *Bijdragen tot de Taal-, Land- en Volkenkunde van Nederlandsch-Indië* 100 (1941), 255-82

Kaye, I., 'Captain James Cook and the Royal Society', *Notes and Records of the Royal Society of London* 24, no. 1 (1969), 7-18

Kehoe, M.L., 'Dutch Batavia: Exposing the Hierarchy of the Dutch Colonial City', *Journal of Historians of Netherlandish Art* 7, no. 1 (2015), 1-35

Keir, B., 'Captain Cook's Longitude Determinations and the Transit of Mercury: Common Assumptions Questioned', *Journal of the Royal Society of New Zealand* 40, no. 2 (2010), 27-38

Kettering, A.M., 'Gentlemen in Satin: Masculine Ideals in Later Seventeenth-Century Dutch Portraiture', *Art Journal* 56, no. 2 (1997), 41-47
Keuning, J., 'Hessel Gerritsz', *Imago Mundi* 6, no. 1 (1949), 48-66
Killingray, D., 'The Maintenance of Law and Order in British Colonial Africa', *African Affairs* 85, no. 340 (1986), 411-37
King, R.J., 'Puerto del Pendulo, Doubtful Sound: The Malaspina Expedition's Visit to New Zealand in Quest of the True Figure of the Earth', *The Globe* 65, no. 1 (2010), 1-18
Knight, T.M., 'Cook the Cartographer', *Cartography* 7, no. 3 (1971), 110-18
Koudijs, P., 'Those Who Know Most: Insider Trading in Eighteenth-Century Amsterdam', *Journal of Political Economy* 123, no. 6 (2015), 1356-1409
Koyoumjian, P., 'Ownership and Use of Maps in England, 1660-1760', *Imago Mundi* 73, no. 1 (2021), 32-45
Krogt, P. van der, 'Selected Papers from the 16th International Conference on the History of Cartography: Amsterdam Atlas Production in the 1630s: A Bibliographer's Nightmare', *Imago Mundi* 48, no. 1 (1996), 149-60
Kuhn, A.J., 'Dr. Johnson Williams, and the Eighteenth-Century Search for the Longitude', *Modern Philology* 82, no. 1 (1984), 40-52
Langdon, R., 'The European Ships of Tupaia's Chart: An Essay in Identification', *Journal of Pacific History* 15, no. 4 (1980), 225-32
Lape, P.V., 'Political Dynamics and Religious Change in the Late Pre-colonial Banda Islands, Eastern Indonesia', *World Archaeology* 32, no. 1 (2000), 138-55
Levtzion, N., 'Ibn-Hawqal, the Cheque, and Awdaghost', *Journal of African History* 9, no. 2 (1968), 223-33
Littlehales, G.W., 'The Decline of the Lunar Distance for the Determination of the Time and Longitude at Sea', *Bulletin of the American Geographical Society* 41, no. 2 (1909), 83-86
Lucassen, J., 'A Multinational and Its Labor Force: The Dutch East India Company, 1595-1795', *International Labor and Working-Class History* 66 (2004), 12-39
Lydon, J., 'Visions of Disaster in the Unlucky Voyage of the Ship *Batavia*, 1647', *Itinerario* 42, no. 3 (2018), 351-74
Maskelyne, N., 'XLIX. Description of a Method of Measuring Differences of Right Ascension and Declination, with Dollond's Micrometer, together with Other New Applications of the Same', *Philosophical Transactions* 61 (1771), 536-46
Massarella, D., 'Chinese, Tartars and "Thea" or a Tale of Two Companies: The English East India Company and Taiwan in the Late Seventeenth Century', *Journal of the Royal Asiatic Society* 3, no. 3 (1993), 393-426
Milligan, R.R.D., 'Ranginui, Captive Chief of Doubtless Bay, 1769', *The Journal of the Polynesian Society* 67, no. 3 (1958), 181-203
Monthly Review, The, or, Literary Journal 40 (London: R. Griffiths, 1769)
Moon, P., 'Entering the Periphery: Reassessing British Involvement in New Zealand in the 1820s in the Context of Wallerstein's Theory of a World System', *New Zealand Journal of History* 49, no. 2 (2015), 81-109

Moon, P., 'From Tasman to Cook: The Proto-intelligence Phase of New Zealand's Colonisation', *Journal of Intelligence History* 18, no. 2 (2019), 253-68
Moon, P., 'Shade of the Savage in Colonial New Zealand', *Journal of Colonialism and Colonial History* 18, no. 2 (2017), 1-17
Moorman, G., 'Publishers at the Intersection of Cultures: The Significance of Italo-Dutch Contacts in the Creation Process of Joan Blaeu's *Theatrum Italiae* (1663)', *Incontri* 30, no. 2 (2015)
Morris, R., '200 Years of Admiralty Charts and Surveys', *The Mariner's Mirror* 82, no. 4 (1996), 420-35
Murphy, P., 'Creating a Commonwealth Intelligence Culture: The View from Central Africa 1945-1965', *Intelligence and National Security* 17, no. 3 (2002), 131-62
Obeyesekere, G., '"British Cannibals": Contemplation of an Event in the Death and Resurrection of James Cook, Explorer', *Critical Inquiry* 18, no. 4 (1992), 630-54
Oliphant, J.O., 'A Project for a Christian Mission on the Northwest Coast of America, 1798', *Pacific Northwest Quarterly* 36, no. 2 (1945), 99-114
Parsons, E.J.S., and W.F. Morris, 'Edward Wright and His Work', *Imago Mundi* 3, no. 1 (1939), 61-71
Pastoureau, M., 'Jacques-Nicolas Bellin, French Hydrographer, and the Royal Society in the Eighteenth Century', *Yale University Library Gazette* 68, nos 1/2 (1993)
Pearson, W.H., 'Hawkesworth's Alterations', *Journal of Pacific History* 7, no. 1 (1972), 45-72
Perrin, W.G., 'The Prime Meridian', *The Mariner's Mirror* 13, no. 2 (1927), 109-24
Peters, M.A., and T. Besley, 'The Royal Society, the Making of "Science" and the Social History of Truth', *Educational Philosophy and Theory* 51, no. 3 (2019), 227-32
Pettegree, A., and A.T. der Weduwen, 'What Was Published in the Seventeenth-Century Dutch Republic?', *Livre. Revue Historique* (2018), 1-27
Pontani, F., 'The World on a Fingernail: An Unknown Byzantine Map, Planudes, and Ptolemy', *Traditio* 65 (2010), 177-200
Raffles, T.S., 'The Founding of Singapore', *Journal of the Straits Branch of the Royal Asiatic Society* 2 (December 1878), 175-82
Raine, K., 'Blake's "Marriage of Heaven and Hell" by Martin K. Nurmi', *The Modern Language Review* 53, no. 2 (1958), 246-48
Raj, K., 'Colonial Encounters and the Forging of New Knowledge and National Identities: Great Britain and India, 1760-1850', *Osiris* 15 (2000), 119-34
Ravneberg, R.L., 'The Hawkesworth Copy: An Investigation into the Printer's Copy Used for the Preparation of the 1773 Second Edition of John Hawkesworth's Account of Captain Cook's First Voyage', *Bibliographical Society of Australia and New Zealand Bulletin* 26, nos 3/4 (2002), 173-92
Richard, H., 'The Interest of French Cartography in the South Pacific in the Eighteenth Century', *The Globe*, 83 (2018), 1-11

Roncière, M. de la, 'Manuscript Charts by John Thornton, Hydrographer of the East India Company (1669-1701)', *Imago Mundi* 19 (1965), 46-50

Rotschi, H., and L. Lemasson, 'Oceanography of the Coral and Tasman Seas', *Oceanogr. Mar. Biol. Ann. Rev.* 5 (1967), 49-98

Saldanha, A., 'The Itineraries of Geography: Jan Huygen van Linschoten's *Itinerario* and Dutch Expeditions to the Indian Ocean, 1594-1602', *Annals of the Association of American Geographers* 101, no. 1 (January 2011), 149-77

Schilder, G., 'Die Entdeckung Australiens im Niederländischen Globusbild des 17 Jahrhunderts', *Der Globusfreund* 25 (1978), 183-94

Schilder, G., 'New Cartographical Contributions to the Coastal Exploration of Australia in the Course of the Seventeenth Century', *Imago Mundi* 26, no. 1 (1972), 41-44

Schilder, G., 'Organization and Evolution of the Dutch East India Company's Hydrographic Office in the Seventeenth Century', *Imago Mundi* 28, no. 1 (1976), 61-78

Schuldt, K., 'Abel Janszoon Tasman', *Deutsches Schiffahrtsarchiv* 8 (1985)

Setiawan, J., and D. Kumalasari, 'The Struggle of Sultan Babullah in Expelling Portuguese from North Maluku', *HISTORIA: Jurnal Pendidik dan Peneliti Sejarah* 2, no. 1 (2019), 1-6

Slack, P., 'Government and Information in Seventeenth-Century England', *Past & Present* 184, no. 1 (2004), 33-68

Slessor, K., 'The Atlas', in A.J. Haft, 'Imagining Space and Time in Kenneth Slessor's "Dutch Seacoast" and Joan Blaeu's Town Atlas of The Netherlands: Maps and Mapping in Kenneth Slessor's Poetic Sequence *The Atlas*, Part Three', *Cartographic Perspectives* 74 (2013), 29-54

Smith, P.M., 'Mapping Australasia', *History Compass* 7, no. 4 (2009), 1099-122

Sorrenson, R., 'The Ship as a Scientific Instrument in the Eighteenth Century', *Osiris* 11 (1996), 221-36

Spencer, J.R.H., 'The New Zealand Charts Compiled on HM Bark Endeavour by Cook, Molyneux and Pickersgill, 1769-70', *Archifacts: Bulletin of the Archives and Records Association of New Zealand*, no. 1 (March 1985), 2-12

Spencer, J.R.H., 'A New Zealand Draught Chart by James Cook, RN', *Cartography* 13, no. 4 (1984), 316-22

Spencer, J.R.H., 'The Visit of the *St Jean-Baptiste* Expedition to Doubtless Bay, New Zealand, 17-31 December 1769', *Cartography* 14, no. 2 (1985), 124-43

Steinberg, P.E., 'Calculating Similitude and Difference: John Seller and the "Placing" of English Subjects in a Global Community of Nations', *Social & Cultural Geography* 7, no. 5 (2006), 687-707

Stimson, A., 'The Influence of the Royal Observatory at Greenwich upon the Design of 17th and 18th Century Angle-Measuring Instruments at Sea', *Vistas in Astronomy* 20 (1976), 123-30

Stokes, E., 'European Discovery of New Zealand before 1642: A Review of the Evidence', *New Zealand Journal of History* 4, no. 1 (1970), 3-19

Stone, J.C., 'Imperialism, Colonialism and Cartography', *Transactions of the Institute of British Geographers* 13, no. 1 (1988), 57-64

Stone, J.C., 'A Newly Discovered Map of Ettrick Forest, Scotland by Robert Gordon of Straloch: Implications for Sources Consulted by Joan Blaeu', *Imago Mundi* 31, no. 1 (1979), 85-87

Taylor, E.G.R., 'Five Centuries of Dead Reckoning', *The Journal of Navigation* 3, no. 3 (1950), 280-85

Taylor, E.G.R., 'Navigation in the Days of Captain Cook', *The Journal of Navigation* 21, no. 3 (1968), 256-76

Tent, J., 'Who Named New Holland?', Occasional Paper no. 13 (Sydney: Australian National Placenames Survey, 2022)

Thomas, M., 'Colonial States as Intelligence States: Security Policing and the Limits of Colonial Rule in France's Muslim Territories, 1920-40', *Journal of Strategic Studies* 28, no. 6 (2005), 1033-60

Tregonning, K., 'Alexander Dalrymple: The Man Whom Cook Replaced', *Australian Quarterly* 23, no. 3 (1951), 54-63

Wallerstein, I., 'The Rise and Future Demise of the World Capitalist System: Concepts for Comparative Analysis', *Comparative Studies in Society and History* 16, no. 4 (September 1974), 387-415

Wallis, H., 'The Eva G.R. Taylor Lecture: Navigators and Mathematical Practitioners in Samuel Pepys's Day', *Journal of Navigation* 47, no. 1 (January 1994), 1-19

Warner, D.J., 'True North: And Why It Mattered in Eighteenth Century America', *Proceedings of the American Philosophical Society* 149, no. 3 (2005), 372-85

Warren, B., 'Maps as Social History (Review)', *Huntington Library Quarterly* 64, nos 1/2 (2001), 275-78, 282

Winterbottom, A., 'Producing and Using the Historical Relation of Ceylon: Robert Knox, the East India Company and the Royal Society', *British Journal for the History of Science* 42, no. 4 (2009), 515-38

Woolley, R., 'Captain Cook and the Transit of Venus of 1769', *Notes and Records of the Royal Society of London* 24, no. 1 (1969)

Zukas, A., 'Class, Imperial Space, and Allegorical Figures of the Continents on Early-Modern World Maps', *Environment, Space, Place* 10, no. 2 (2018), 29-62

Books and Chapters in Books

Adams, P.G., *Travel Literature and the Evolution of the Novel* (Lexington: University Press of Kentucky, 1983)

Anderson, G., *The Merchant of Zeehaen: Isaac Gilsemans and the Voyages of Abel Tasman* (Wellington: Te Papa Press, 2001)

Anson, G., *A Voyage Round the World in the Years MDCCXL, I, II, III, IV* (London: John and Paul Knapton, 1748)

Anson, G., *A Voyage Round the World in the Years MDCCXL, I, II, III, IV* (London: H. Woodfall, 1769)

Armitage, D., *The Ideological Origins of the British Empire* (Cambridge: Cambridge University Press, 2000)

Ballantyne, T., *Webs of Empire: Locating New Zealand's Colonial Past* (Wellington: Bridget Williams Books, 2012)
Banks, J., *The Endeavour Journal of Joseph Banks, 1768-1771*, ed. J.C. Beaglehole, 2 vols (Sydney: Angus and Robertson Ltd, 1962)
Barber, P., *The Map Book* (London: Weidenfeld & Nicolson, 2005)
Barnard, H., 'Maps and Mapmaking in Ancient Egypt', in H. Selin (ed.), *Encyclopaedia of the History of Science, Technology, and Medicine in Non-Western Cultures* (Berlin: Springer-Verlag, 2008), pp. 1273-76
Bartholomew, T.A., *Richard Bentley, D.D.: A Bibliography of His Works and of All the Literature Called Forth by His Acts or His Writings* (Cambridge: Bowes & Bowes, 1908)
Bayly, C., *Empire and Information: Intelligence Gathering and Social Communication in India, 1780-1870* (Cambridge: Cambridge University Press, 1996)
Berggren, J.L., and A. Jones, *Ptolemy's Geography: An Annotated Translation of the Theoretical Chapters* (Princeton: Princeton University Press, 2001)
Birdwood, G.C.M., *Report on the Old Records of the India Office, with Supplementary Note and Appendices* (London, W.H. Allen & Co., 1891)
Blaeu, W.J., *Atlantis majoris appendix sive pars altera continens geographicas tabulas diversarum orbis regionum et provinciarum octoginta* (Amsterdam, Willem Janszoon Blaeu, 1630)
Briggs, A., and P. Burke, *A Social History of the Media from Gutenberg to the Internet* (Cambridge: Polity Press, 2005)
Brosses, C. de, *Histoire des navigations aux terres australes* (Paris: Chez Durand, 1756)
Buning, M., 'Privileging the Common Good: The Moral Economy of Printing Privileges in the Seventeenth-Century Dutch Republic', in S. Graheli (ed.), *Buying and Selling: The Business of Books in Early Modern Europe* (Leiden: Brill, 2019), pp. 88-108
Burke, E., *A Vindication of Natural Society: or, A View of the Miseries and Evils Arising to Mankind* (London: M. Cooper, 1756)
Burnet, I., *The Tasman Map: The Biography of a Map* (Sydney: Rosenberg Publishing, 2019)
Burney, F., *The Early Journals and Letters of Fanny Burney: Volume 1: 1768-1773*, ed. L.E. Troide (Kingston, ON: McGill-Queen's University Press, 1988)
Burney, J., Capt., *A Chronological History of the Discoveries in the South Sea or Pacific Ocean*, 5 vols (London: Luke Hansard, 1803-17)
Burney, J., Capt., *A Chronological History of the Voyages and Discoveries in the South Sea or Pacific Ocean: Part III: From the Year 1620 to the Year 1688: Illustrated with Charts and Other Plates* (London: Luke Hansard & Sons, 1813)
Callander, J., and C. de Brosses, *Terra Australis Cognita, or, Voyages to the Terra Australis, or Southern Hemisphere, During the Sixteenth, Seventeenth, and Eighteenth Centuries* (Edinburgh: J. Donaldson, 1768)
Campbell, J., *A Political Survey of Britain: Being a Series of Reflections on the Situation, Lands, Inhabitants, Revenues, Colonies, and Commerce of this Island* (London: Richardson & Urquhart, 1774), Vol. 1

Carpenter, K.J., *The History of Scurvy and Vitamin C* (Cambridge: Cambridge University Press, 1988)
Carrington, H., (ed.), *The Discovery of Tahiti, A Journal of the Second Voyage of HMS Dolphin Round the World, under the Command of Captain Wallis, RN: In the Years 1766, 1767, and 1768, Written by her Master, George Robertson* (London: Hakluyt Society, 2017)
Castells, M., *The Information Age: Economy, Society and Culture: Volume 1: The Rise of the Network Society* (Oxford: Blackwell, 1996)
Castells, M., *The Informational City: Information, Technology, Economic Restructuring, and the Urban-Regional Process* (Oxford: Basil Blackwell, 1989)
Charry, B., *The Tempest: Language and Writing* (London: Bloomsbury, 2013)
Chaudhuri, K.N., *The English East India Company: The Study of an Early Joint-Stock Company 1600-1640* (London: Frank Cass & Co., 1965)
Childs, E.C., *Vanishing Paradise: Art and Exoticism in Colonial Tahiti* (Berkeley: University of California Press, 2013)
Conrad, J., *Heart of Darkness* (Oxford: Oxford University Press, 2002)
Cook, J., *Directions for Navigating the West-Coast of Newfoundland, with a Chart Thereof* (London: J. Mount & T. Page, 1766)
Cook, J., *The Journals of Captain James Cook on His Voyages of Discovery*, ed. J.C. Beaglehole, 4 vols plus portfolio (Cambridge: Hakluyt Society, 1955)
Coreal, F., *Voyages de François Coreal aux Indes Occidentales, contenant ce qu'il y a vû de plus remarquable pendant son séjour depuis 1666 jusqu'en 1697: traduits de l'espagnol. Avec une relation de la Guiane de Walter Raleigh, & le voyage de Narbrough à la mer de Sud par le détroit de Magellan, &c* (Paris: Chez André Cailleau, 1722)
Crozet, J.-M., *Nouveau voyage à la mer du Sud commencé sous les ordres de M. Marion* (Paris: Barrois, 1783)
Curnow, A., 'Landfall in Unknown Seas', in A. Curnow, *Collected Poems, 1933-1973* (Wellington: A.H. & A.W. Reed, 1974)
Dalrymple, A., *An Account of What Has Passed between the India Directors and Alexander Dalrymple* (London: J. Nourse 1769)
Dalrymple, A., *An Historical Collection of the Several Voyages and Discoveries in the South Pacific Ocean*, 2 vols (London: J. Nourse, 1770, 1771)
Dam, P. van, *Beschryvinge van de Oostindische Compagnie*, ed. F.W. Stapel (The Hague: Martinus Nijhoff, 1927-54), Vol. 1
Dampier, W., *Voyages and Descriptions* (London: James Knapton, 1700), Vol. 2
David, A., (ed.), *The Charts and Coastal Views of Captain Cook's Voyages: Volume 1: The Voyage of the* Endeavour, *1768-1771* (London: Hakluyt Society, 1988)
Derham, W., *Physico-Theology: or, A Demonstration of the Being and Attributes of God* (London: W. Innys, 1713)
Destombes, M., *Cartes Hollandaises: La cartographie de la Compagnie des Indes Orientales, 1593-1743* (Saigon: C. Ardin, 1941)
Dickinson, H.T., *The Politics of the People in Eighteenth-Century Britain* (Basingstoke: Macmillan, 1994)

Dolan, G., *The Quest for Longitude and the Rise of Greenwich: A Brief History* (London: The Royal Observatory Greenwich, 2022)
Drake-Brockman, H., *Voyage to Disaster: The Life of Francisco Pelsaert Covering His Indian Report to the Dutch East India Company and the Wreck of the Ship* Batavia *in 1629* (London: Angus & Robertson, 1982)
Dunmore, J., (ed.), *The Expedition of the* St Jean-Baptiste *to the Pacific, 1769-1770* (London: Hakluyt Society, 1981)
Eade, J., and M.J. Sallnow (eds), *Contesting the Sacred: The Anthropology of Christian Pilgrimage* (London: Routledge, 1991)
Edmond, M., *Zone of the Marvellous: In Search of the Antipodes* (Auckland: Auckland University Press, 2009)
Edney, M.H., 'The Irony of Imperial Mapping', in J.R. Akerman (ed.), *The Imperial Map: Cartography and the Mastery of Empire* (Chicago: University of Chicago Press, 2009)
Edney, M.H., *Mapping an Empire: The Geographical Construction of British India, 1765-1843* (Chicago: University of Chicago Press, 2009)
Eisler, W., *The Furthest Shore: Images of Terra Australis from the Middle Ages to Captain Cook* (Cambridge: Cambridge University Press, 1995)
Enthoven, V., *Zeeland en de opkomst van de Republiek: Handel en strijd in de Scheldedelta,* c. *1550-1621* (Leiden: Luctor et Victor, 1999)
Enthoven, V., S. Murdoch and E. Williamson (eds), *The Navigator: The Log of John Anderson, VOC Pilot-Major, 1640-1643* (Leiden: Brill, 2010)
Forbes, E., *Greenwich Observatory: The Royal Observatory at Greenwich and Herstmonceux: Volume 1: Origins and Early History (1675-1835)* (London: Taylor & Francis, 1975)
Forster, G., *A Voyage Round the World, in His Britannic Majesty's Sloop, Resolution, Commanded by Capt. James Cook, during the Years 1772, 3, 4, and 5*, 2 vols (London: G. Robinson, 1778)
Foster, W., *The East India House: Its History and Associations* (London: John Lane, 1924)
Frank, A.G., *Latin America: Underdevelopment or Revolution: Essays on the Development of Underdevelopment and the Immediate Enemy* (New York: Monthly Review Press, 1970)
Fry, H.T., 'Alexander Dalrymple and Captain Cook: The Creative Interplay of Two Careers', in R. Fisher and H. Johnston (eds), *Captain James Cook and His Times* (Canberra: Australian National University Press, 1979)
Gaastra, F.S., *Geschiedenis van de VOC: Opkomst, bloei en ondergang* (Amsterdam: Amsterdam University Press, 2016)
Gerritsz, H., *History of the Country Called Spitsbergen: Its Discovery, Its Situation, Its Animals* [Amsterdam, 1613], trans. (London: British Museum, 1902)
Gilmour, D., *The Ruling Caste: Imperial Lives in the Victorian Raj* (London: Pimlico, 2005)
Gleig, G.R. (ed.), *Memoirs of the Life of the Right Hon. Warren Hastings* (London: R. Bentley, 1841), Vol. 1

Godson, W., *A New and Correct Map of the World: Laid down according to the newest observations & discoveries in several different projections including the trade winds, monsoons, variation of the compass, and illustrated with a coelestial planisphere, the various systems of Ptolomy, Copernicus, and Tycho Brahe together with ye apearances of the planets &c.* (London: George Willdey, 1702)

Goossens, E.-J., *Treasure Wrought by Chisel and Brush: The Town Hall of Amsterdam in the Golden Age* (Zwolle: Waanders, 1996)

Greene, J.P., 'Empire and Identity from the Glorious Revolution to the American Revolution', in P.J. Marshall (ed.), *The Oxford History of the British Empire: The Eighteenth Century* (Oxford: Oxford University Press, 1998)

Harris, J., *Navigantium atque Itinerantium Bibliotheca* (London: Thomas Bennet, 1705)

Harris, J., *Navigantium atque Itinerantium Bibliotheca, or, a Complete Collection of Voyages and Travels*, ed. John Campbell (London: T. Woodward, 1744)

Hartkamp-Jonxis, E., *Sits: Oost-West Relaties in Textiel* (Zwolle: Waanders, 1987)

Hawkesworth, J., *An Account of the Voyages Undertaken by the Order of His Present Majesty for Making Discoveries in the Southern Hemisphere*, 3 vols (London: W. Strahan & T. Cadell, 1773)

Hawkesworth, J., *Relation des voyages entrepris par ordre de sa Majesté Britannique, actuellement regnante; Pour faire des découvertes dans l'hémisphère méridional, Tome III* (Paris: Saillant et Nyon, 1774)

Heeres, J.E., and C.H. Coote (eds), *Abel Janszoon Tasman's Journal* (Los Angeles: Kovach, 1965)

Henry, R., (ed.), *Early Voyages to Terra Australis, Now Called Australia: A Collection of Documents, and Extracts from Early Manuscript Maps* (London: Hakluyt Society, 1859)

Herodotus, *'On Libya', from The Histories, c. 430 BC*, text provided by Prof. J.S. Arkenberg of California State University, Fullerton; available online at the Ancient History Sourcebook, History Department, Fordham University, New York (1998)

Heylyn, P., *Cosmographie, in Four Books. Containing the Chorographie and Historie of the Whole World* (London: Henry Seile, 1657)

Historical Records of New South Wales: Volume 1, Part 1: Cooke 1762-1780 (Sydney: Charles Potter, 1893)

Hooke, R., *A Short Relation out of the Journal of Captain Abel Jansen Tasman, upon the Discovery of the South Terra Incognita: not long since Published in the Low Dutch by Dirk Rembrantse* (London: R. Chiswell, 1682)

Hoving, A., and C. Emke, *The Ships of Abel Tasman* (Hilversum: Uitgeverij Verloren, 2000)

Hull, C.H., (ed.), *Economic Writings of Sir William Petty*, 2 vols (Cambridge: Cambridge University Press, 1899)

Hume, D., *The History of England from the Invasion of Julius Caesar to the Revolution in 1688* (London: J. Mcreery, 1807)

Huttich, J., and S. Grynaeus, *Novus orbis regionum ac insularum veteribus incognitarum* (Basel: Hervagius, 1532)
Israel, J., *Dutch Primacy in World Trade, 1585-1740* (Oxford: Clarendon Press, 1989)
Klencke, J., (ed.), *Soli Britannico Reduci Carolo Secundo regum augustissimo hoc Orbis Terrae Compendium* (Amsterdam: Visscher, Blaeu and others, 1660)
Krogt, P. van der, *Koeman's Atlantes Neerlandici New Edition: Volume 2: The Folio Atlases Published by Willem Jansz. Blaeu and Joan Blaeu* (Utrecht: Hes & De Graaf, 2000)
Lalande, J., *Mappemonde* (Paris: n.p, 1762)
Lamb, C., 'The Superannuated Man', in C. Lamb, *Essays of Elia* (London: Macmillan & Co., 1883)
Lane, K.E., *Pillaging the Empire: Piracy in the Americas, 1500-1750* (New York: M.E. Sharpe, 1998)
Langer, W.L., *The Diplomacy of Imperialism, 1890-1902* (New York: Alfred A. Knopf, 1965)
Latham, R., (ed.), *Samuel Pepys and the Second Dutch War: Pepys's Navy White Book and Brooke House Papers* (Oxford: Routledge, 2019)
Latham, R., (ed.), *The Shorter Pepys* (London: Bell & Hyman, 1985)
Lawrence, D.H., *Sea and Sardinia* (New York: Thomas Seltzer, 1921)
Lee, M., *Navigators & Naturalists: French Exploration of the New Zealand and the South Pacific Seas (1769-1824)* (Auckland: David Bateman, 2018)
Leerssen, J., *National Thought in Europe: A Cultural History* (Amsterdam: Amsterdam University Press, 2006)
Linnaeus, C., *Systema naturae per regna tria naturae* (Stockholm: Laurentii Salvii, 1758)
Linschoten, J.H. van, *Itinerario: Voyage ofte Schipvaert naer Oost ofte Portugaels Indien* (Amsterdam: Cornelis Claesz, 1596)
Lockett, J., *Captain James Cook in Atlantic Canada* (Halifax: Formac Publishing, 2010)
Long, D.A., *'At the Sign of Atlas': The Life and Work of Joseph Moxon, A Restoration Polymath* (Donington: Shaun Tyas, 2013)
Lyons, H.G., *The Royal Society, 1660-1940: A History of Its Administration under Its Charters* (Cambridge: Cambridge University Press, 1944)
Macartney, G., *An Account of Ireland in 1773: By a Late Chief Secretary of that Kingdom* (London, n.p., 1773)
Macgregor, J., *Commercial Statistics: A Digest of the Productive Resources, Commercial Legislation, Customs Tariffs, of All Nations. Including All British Commercial Treaties with Foreign States* (London: Whittaker & Co., 1850), Vol. 3
Markham, C., (ed.), *The Voyages of Pedro Fernandez de Quiros*, 1595-1606, 2 vols (London: Hakluyt Society, 1904)
Marra, J., *Journal of the Resolution's Voyage: in 1772, 1773, 1774, and 1775* (Dublin: Caleb Jenkin, 1776)

Marshall, P.J., (ed.), *The Oxford History of the British Empire: Volume 2: The Eighteenth Century* (Oxford: Oxford University Press, 1998)
Maskelyne, N., *The Nautical Almanac and Astronomical Ephemeris* (London: J. Nourse, 1766)
McCormick, E.H., *Tasman and New Zealand: A Bibliographical Study* (Bulletin no. 14) (Wellington: Government Printer, 1959)
McCulloch, J.R., *A Treatise on the Principles, Practice, & History of Commerce* (London: Baldwin & Cradock, 1833)
McNab, R., *From Tasman to Marsden: A History of Northern New Zealand from 1642 to 1818* (Dunedin: J. Wilkie & Co., 1914)
McNab, R., (ed.), *Historical Record of New Zealand: Volume 2* (Wellington: John Mackay, 1914)
McNab, R., *Murihiku and the Southern Islands: A History of the West Coast Sounds, Foveaux Strait, Stewart Island, the Snares, Bounty, Antipodes, Auckland, Campbell and Macquarie Islands from 1770 to 1829* (Invercargill: William Smith, 1907)
Mercator, G., *Atlas ou Representation du Monde Universel* (Amsterdam: Chez Henry Hondius, 1633)
Meyjes, R., (ed.), *De Reizen van Abel Janszoon Tasman en Franchoys Jacobszoon Visscher Ter Nadere Ontdekking von Het Zuidland in 1642/3 en 1644* (The Hague: Martinus Nijhoff, 1919)
Milburn, W., *Oriental Commerce: Containing a Geographical Description of the Principal Places in the East Indies, China, and Japan* (London: Black, Parry & Co., 1813)
Morris, M., (ed.), *Boswell's Life of Johnson* (London: Macmillan & Co., 1899)
Moxon, J., *A Brief Discourse of a Passage by the North Pole to Japan, China. &c* (London: Joseph Moxon, 1674)
Moxon, J., *Regulae Trium Ordinum Literarum Typographicarum, or, The Rules of the Three Orders of Print Letters* (London: Joseph Moxon, 1676)
Narborough, J., *An Account of Several Late Voyages and Discoveries to the South and North* (London: Sam Smith, 1694)
National Library of Australia, *Mapping Our World: Terra Incognita to Australia* (Canberra: National Library of Australia, 2013)
Nautical Almanac, The, and Astronomical Ephemeris, for the Year 1767 (London: W. Richardson & S. Clark, 1766)
Nellen, H., and P. Steenbakkers, 'Biblical Philology in the Long Seventeenth Century: New Orientations', in D. van Miert, H. Nellen, P. Steenbakkers and J. Touber (eds), *Scriptural Authority and Biblical Criticism in the Dutch Golden Age: God's Word Questioned* (Oxford: Oxford University Press, 2017)
Nette, D. van, 'The New World Map and the Old: The Moving Narrative of Joan Blaeu's Nova Totius Terrarum Orbis Tabula (1648)', in B. Vannieuwenhuyze and Z. Segal (eds.), *Motion in Maps, Maps in Motion: Mapping Stories and Movement through Time* (Amsterdam: Amsterdam University Press, 2020)

Nietzsche, F., *The Will to Power* (New York: Vintage Books, 1968)
Ogilby, J., *America: Being an Accurate Description of the New World; Containing the Original of the Inhabitants; the Remarkable Voyages Thither* (London: Thomas Johnson, 1670)
Overton, H., *A New & Correct Map of the Trading Part of the West Indies, including the Seat of War between Gr. Britain and Spain: Likewise the British Empire in America, with the French and Spanish Settlements Adjacent* (London: Henry Overton, at the White Horse without Newgate, 1741)
Parkinson, S., *A Journal of a Voyage to the South Seas, in His Majesty's Ship, the Endeavour* (London: Stanfield Parkinson, 1773)
Parry, J.H., *The Age of Reconnaissance: Discovery, Exploration and Settlement, 1450-1650* (London: Phoenix Press, 1962)
Paul, S., *Jeopardy of Every Wind: The Biography of Captain Thomas Bowrey* (London: Monsoon Books, 2020)
Pedley, M.S., *The Commerce of Cartography: Making and Marketing Maps in Eighteenth-Century France and England* (Chicago: University of Chicago Press, 2005)
Pepys, S., *The Diary of Samuel Pepys, MA, FRS, 1663* (Frankfurt: Outlook, 2018)
Plato, *Timaeus and Critias*, trans. R. Waterfield (Oxford: Oxford University Press, 2008)
Polack, J., *New Zealand: Being a Narrative of Travels and Adventures during a Residence in that Country*, 2 vols (London: Richard Bentley, 1838)
Porter, T., *A New Booke of Mapps* (London: Robert Walton, 1655)
Prévost, A., *Histoire générale des voyages, ou, Nouvelle collection de toutes les relations de voyages par mer et par terre ... composées sur les observations les plus autentiques* (Paris: Didot, 1746), Vol. 6
Price, A.G., (ed.), *The Explorations of Captain James Cook in the Pacific* (New York: Dover Publications, 1971)
Ray, J., *The Wisdom of God Manifested in the Works of the Creation* (London: R. Harbin, 1717)
Ross, J., *This Stern Coast: The Story of the Charting of the New Zealand Coast* (Wellington: A.H. & A.W. Reed, 1969)
Royal Society, The, *An Alphabetical Catalogue Abreviated in the Philosophical Transactions* (London: Sam Smith, 1694)
Rymsdyk, J. van, and A. van Rymsdyk, *Museum Britannicum: Being an Exhibition of a Great Variety of Antiquities and Natural Curiosities, Belonging to that Noble and Magnificent Cabinet, the British Museum* (London: I. Moore, 1778)
Salmond, A., *Aphrodite's Island: The European Discovery of Tahiti* (Berkeley: University of California Press, 2009)
Salmond, A., *Two Worlds: First Meetings Between Maori and Europeans, 1642-1772* (Auckland: Penguin, 2018)
Sanders, N.K., *The Epic of Gilgamesh* (London: Penguin, 2006)
Scammell, G.V., *The First Imperial Age: European Overseas Expansion* c. *1400-1715* (London: Routledge, 2003)

Schama, S., *The Embarrassment of Riches: An Interpretation of Dutch Culture in the Golden Age* (New York: Vintage Books, 1997)
Schama, S., *A History of Britain: Volume 2: The British Wars 1603-1776* (London: The Bodley Head, 2009)
Schilder, G., *Australia Unveiled: The Share of the Dutch Navigators in the Discovery of Australia*, trans. O. Richter (Amsterdam: Theatrum Orbis Terrarum, 1976)
Schilder, G., *Monumenta Cartographica Neerlandica*, 9 vols (Alphen aan den Rijn: Canaletto, 2000)
Schouten, W., *Journal ou description du merveilleux voyage de Guillaume Schouten, Hollandois natif de Hoorn, fait es années 1615, 1616 & 1617* (Amsterdam: Harman Janson, 1619)
Shapin, S., 'Who was Robert Hooke?', in M. Hunter and S. Schaffer (eds.), *Robert Hooke: New Studies* (Woodbridge, Suffolk: The Boydell Press, 1989), pp. 253-85
Sharp, A., *The Voyages of Abel Janszoon Tasman* (Oxford: Clarendon Press, 1968)
Slot, B.J., *Abel Tasman and the Discovery of New Zealand* (Amsterdam: Otto Cramwinckel, 1992)
Smith, S., *An Account of Several Late Voyages and Discoveries to the South and North* (London: Sam Smith, 1694)
Spate, O.H.K., *Monopolists and Freebooters [The Pacific Since Magellan: Volume 2]* (Canberra: Australian National University Press, 1983)
Steele, I.K., *The English Atlantic, 1675-1740: An Exploration of Communication and Community* (Oxford: Oxford University Press, 1986)
Steensgaard, N., 'The Dutch East India Company as an Institutional Innovation', in M. Aymard (ed.), *Dutch Capitalism and World Capitalism* (Cambridge: Cambridge University Press, 1982)
Stevens, H., *The Dawn of British Trade to the East Indies as Recorded in the Court Minutes of the East India Company 1599-1603* (London: Henry Stevens & Son, 1886)
Stronks, E., 'The Diffusion of Illustrated Religious Texts and Ideological Restraints', in A. Lardinois, S. Levie, H. Hoeken and C. Lüthy (eds), *Texts, Transmissions, Receptions: Modern Approaches to Narratives* (Leiden: Brill, 2015), pp. 194-220
Suarez, T., *Early Mapping of Southeast Asia* (Hong Kong: Periplus, 1999)
Surville, J.F.M. de, *Extracts from Journals Relating to the Visit to New Zealand of the French Ship St Jean Baptiste in December 1769 Under the Command of JFM de Surville*, trans. I. Ollivier and C. Hingley (Wellington: Alexander Turnbull Library Endowment Trust, 1982)
Taylor, E.G.R., *The Mathematical Practitioners of Hanoverian England, 1714-1840* (Cambridge: Cambridge University Press, 1966)
Taylor, J.G., *The Social World of Batavia: European and Eurasian in Dutch Asia* (Wisconsin: University of Wisconsin Press, 2004)
Thomson, J., *Alfred: A Masque* (London: A. Millar, 1751)

Thornton, J., *Atlas Maritimus or, The Sea-Atlas: Being a Book of Maritime Charts* (London: n.p., 1700)
Tooley, R.V., *Maps and Mapmakers*, 6th edn (London: Batsford, 1978)
Valentijn, F., *Oud en Nieuw Oost-Indiën, vervattende een naaukeurige en uitvoerige verhandelinge van Nederlands mogentheyd in die gewesten, benevens eene wydluftige beschryvinge der Moluccos, Amboina, Banda, Timor, en Solor, Java, en alle de eylanden onder dezelve landbestieringen behoorende, het Nederlands comptoir op Suratte, en de levens der Groote Mogols* (Dordrecht: By Joannes van Braam, Gerard Onder de Linden, 1724-26)
Vancouver, G., *A Voyage of Discovery to the North Pacific Ocean, and Round the World ... Performed in the Years 1790, 1791, 1792, 1793, 1794 and 1795* (London: John Stockdale, 1801)
Vespucci, A., *Mundus Novus: Letter to Lorenzo Pietro di Medici* [Vienna, 1504], trans. G.T. Northup (Princeton: Princeton University Press, 1916)
Vries, J. de, 'Understanding Eurasian Trade in the Era of the Trading Companies', in M. Berg (ed.), *Goods from the East, 1600-1800: Trading Eurasia* (London: Palgrave Macmillan, 2015)
Vries, J. de, and A. van der Woude, *The First Modern Economy: Success, Failure, and Perseverance of the Dutch Economy, 1500-1815* (Cambridge: Cambridge University Press, 1997)
Vulliamy, C.E., *James Boswell* (London: Geoffrey Bles, 1932)
Walker, J.B., *Abel Janszoon Tasman: His Life and Voyages: Read before the Royal Society of Tasmania, 25th November, 1895* (Hobart: Government Printer, 1896)
Waller, R., *The Posthumous Works of Dr. Robert Hooke* (London: Sam Smith, 1705)
Weld, C.R., *A History of the Royal Society, with Memoirs of the Presidents*, 2 vols (London: John W. Parker, 1848)
Whitfield, P., *The Image of the World: 20 Centuries of World Maps* (San Francisco: Pomegranate Artbooks, 1994)
Wieder, F.C., *Monumenta Cartographica: Reproductions of Unique and Rare Maps* (The Hague: Martinus Nijhoff, 1925-33), Vol. 3
Williams, G., 'The Pacific: Exploration and Exploitation', in P.J. Marshall (ed.) *The Oxford History of the British Empire: Volume 2: The Eighteenth Century* (Oxford: Oxford University Press, 1998), pp. 552-75
Williamson, G., *British Masculinity in* The Gentleman's Magazine, *1731 to 1815* (Basingstoke: Palgrave Macmillan, 2016)
Wilson, K., 'Empire of Virtue: The Imperial Project and Hanoverian Culture *c*. 1720-1785', in L. Stone (ed.), *An Imperial State at War: Britain from 1689 to 1815* (London: Routledge, 1994)
Wilson, R., Voyages of Discoveries Round the World (London: James Cundee, 1806), Vol. 1
Withey, L., *Voyages of Discovery: Captain Cook and the Exploration of the Pacific* (Berkeley: University of California Press, 1989)

Woods, M., 'For the Dutch Republic, the Great Pacific', in National Library of Australia, *Mapping our World: Terra Incognita to Australia* (Canberra: National Library of Australia, 2013)
Woodward, D., (ed.), *The History of Cartography: Volume 3 (Part 2): Cartography in the European Renaissance* (Chicago: University of Chicago Press, 2007)
Wright, E., *Certaine Errors in Navigation* (London: Valentine Sims, 1599)
Wright, E., and J. Moxon, *Certaine Errors in Navigation Detected and Corrected, with Many Additions that were not in the Former Editions* (London: Joseph Moxon, 1657)
Wroth, L.C., *The Early Cartography of the Pacific* (New York: Bibliographical Society of America, 1944)
Zandvliet, K., *De groote waereld in 't kleen geschildert: Nederlandse kartografie tussen de middeleeuwen en de industriële revolutie* (Alphen aan den Rijn: Canaletto, 1985)
Zandvliet, K., 'Golden Opportunities in Geopolitics: Cartography and the Dutch East India Company during the Lifetime of Abel Tasman', in W. Eisler and B. Smith (eds), *Terra Australis: The Furthest Shore* (Sydney: International Cultural Corporation of Australia, 1988), pp. 67-84
Zandvliet, K., 'Mapping the Dutch World Overseas in the Seventeenth Century', in D. Woodward (ed.), *The History of Cartography: Volume 3 (Part 2): Cartography in the European Renaissance* (Chicago: University of Chicago Press, 2007)
Zumthor, P., *Daily Life in Rembrandt's Holland* (London: Weidenfeld & Nicolson, 1962)

Index

You may also be interested in:

Of Seas and Ships and Scientists

The Remarkable History of the UK's National Institute of Oceanography, 1949-1973

Edited by Anthony Laughton, John Gould, 'Tom' Tucker and Howard Roe

This exciting account of a formative phase of UK science begins with scientists working at Antarctic whaling stations prior to World War Two, exploring the biology of the vast icy Southern Ocean in Captain Scott's *Discovery.* During the war, a small group of young scientists was brought together under the inspirational leadership of Dr (later Sir) George Deacon, to study how the movements of the waves affected amphibious landings. These groups were to form the core of the UK's first National Institute of Oceanography when it was founded in 1949. NIO's discoveries in the 1950s and 1960s underpin much of our modern-day marine science, including the complex connections between oceans and climate.

Written by NIO scientists who took part in this crucial period in the development of British marine science, *Of Seas and Ships and Scientists* is the facinating story behind the discoveries which changed our preconceptions of how the oceans 'work'. Their accounts convey the atmosphere of working at sea in an age before portable computers and satellite navigation.

'This is an attractive, well-produced book that will appeal to a wide readership.' – **Gwyn Griffiths, *International Journal of the Society for Underwater technology***

Sir Anthony Laughton joined the NIO in 1955 and was Director of IOS (the successor of NIO) between 1978-88. **Dr John Gould** joined the NIO in 1967 and was head of Marine Physics and later directed two global climate projects. **Mr M.J. 'Tom' Tucker** joined the Group W at the Admiralty Research Laboratory in 1944. He later moved to the NIO. **Prof Howard Roe** joined the Whale Research Unit at the NIO in 1965, following a degree in Zoology at University College London.

Published 2010

Paperback ISBN: 978 0 7188 9230 2
PDF ISBN: 978 0 7188 9703 1
ePub ISBN: 978 0 7188 9702 4

You may also be interested in:

Forging Modernity

Why and How Britain Developed the Industrial Revolution

By Martin Hutchinson

The Industrial Revolution provided the greatest increase in living standards the world has ever known while propelling Britain to dominance on the global stage. In *Forging Modernity*, Martin Hutchinson looks at how and why Britain gained this prize ahead of its European competitors. After comparing their endowments and political structures as far back as 1600, he then traces how Britain, through better policies primarily from the political Tory party, diverged from other European countries. Many early successes resulted from marketing, control systems and logistics rather than from production technology alone, while on a national scale the scientific method and commercial competition were as important as physical infrastructure.

By 1830, through ever-improving policies, Britain had built a staggering industrial lead, half a century ahead of its rivals. In his conclusion, Hutchinson shows how subsequent changes welcomed by conventional historians caused the decline of Industrial Britain. However, the policies that drove growth and rising living standards are still available for those bold enough to adopt them.

'It is refreshing to be told that there is a lot good about economic growth, freeing people from poverty and extending opportunities for many to earn and save more and enjoy a better lifestyle..' – **Rt Hon. Sir John Redwood, MP**

Martin Hutchinson was born in London, and now lives in Poughkeepsie, NY. He was a merchant banker for more than twenty-five years before moving into financial journalism in 2000. He earned his undergraduate degree in mathematics from Trinity College, Cambridge, and an MBA from Harvard Business School. He is also the author of *Britain's Greatest Prime Minister: Lord Liverpool* (Lutterworth Press, 2020).

Published 2023

Hardback ISBN: 978 0 7188 9686 7

Paperback ISBN: 978 0 7188 9689 8

PDF ISBN: 978 0 7188 9687 4

ePub ISBN: 978 0 7188 9688 1

BV - #0103 - 190424 - C0 - 234/156/14 - PB - 9780718897208 - Gloss Lamination